Fatigue Limit in Metals

FOCUS SERIES

Series Editor Gilles Pijaudier-Cabot

Fatigue Limit
in Metals

Claude Bathias

WILEY

First published 2014 in Great Britain and the United States by ISTE Ltd and John Wiley & Sons, Inc.

ISTE Ltd
27-37 St George's Road
London SW19 4EU
UK

www.iste.co.uk

John Wiley & Sons, Inc.
111 River Street
Hoboken, NJ 07030
USA

www.wiley.com

Library of Congress Control Number: 2013950132

British Library Cataloguing-in-Publication Data
A CIP record for this book is available from the British Library
ISSN: 2051-2481 (Print)
ISSN: 2051-249X (Online)
ISBN: 978-1-84821-476-7

MIX
Paper from
responsible sources
FSC
www.fsc.org   FSC® C013604

Printed and bound in Great Britain by CPI Group (UK) Ltd., Croydon, Surrey CR0 4YY

Contents

Acknowledgments

The author is grateful to Professor Paul. C. Paris for his constant encouragement, for his participation in several cooperative research projects on gigacycle fatigue and for his writing of several papers included in this book.

I wish to acknowledge the assistance of colleagues and friends: S. Antolovich (Georgia Tech), H. Mughrabi (Erlangen University), T. Palin-Luc (Ecole Nationale des Arts et Métiers (ENSAM)) and P. Herve (University Paris ouest).

I wish to express special thanks to some PhD students of the University of Paris, who were involved in gigacycle fatigue research during the 2000s: I. Marines and R. Perez Mora (from Mexico), H. Xue, Z. Huang, Weiwei Du and Chong Wang (from China), and A. Nikitin (from Russia).

This work was supported by several financial sources. The most important among them are Ascometal, Safran, Renault, A2MI, Sandvik, Vallourec, Hansen and the French Agency for Research (*L'Agence Nationale de la Recherche* – ANR).

Introduction on Very High Cycle Fatigue

This chapter is a summary of several decades of reasearch on gigacycle fatigue of metals. For more detail please see references [BAT 04] and [BAT 10].

1.1. Fatigue limit, endurance limit and fatigue strength

Fatigue limit, endurance limit and fatigue strength are all expressions used to describe a property of materials under cyclic loading: the amplitude (or range) of *cyclic stress* that can be applied to the material without causing *fatigue failure.* In these cases, a number of cycles (usually 10^7) are chosen to represent the fatigue life of the material.

According to the American Society for Testing and Materials (ASTM) Standard E 1150, the definition of *fatigue* is summarized as follows: "The process of progressive localized permanent structural damage occurring in a material subjected to conditions that produce fluctuating stresses and strains at some point or points and that may culminate in cracks or complete fracture after a sufficient number of fluctuations". The plastic strain resulting from cyclic stress initiates the crack; the tensile stress promotes crack growth propagation. Microscopic plastic strains also can be present at low levels of stress where the strain might otherwise appear to be totally elastic. The ASTM defines *fatigue strength,* S_{Nf}, as the value of stress at which failure occurs after N_f cycles, and *fatigue limit*, S_f, as the limiting value of stress at which failure occurs as N_f becomes very large. The ASTM does not define *endurance limit*, the stress value below which the material will withstand many load cycles, but implies that it is similar to fatigue limit.

Some authors use *endurance limit* for the stress below which failure never occurs, even for an indefinitely large number of loading cycles, as in the case of steel, and *fatigue limit* or *fatigue strength* for the stress at which failure occurs after a specified number of loading cycles, such as 500 million, as in the case of aluminum. Other authors do not differentiate between the expressions even if they do differentiate between the face center cubic (FCC) metals and the base center cubic (BCC) metals [BAT 10].

Since the word "fatigue" was used by Braithwaite, A. Wöhler established the first basic approach to the fatigue life of metals, in the mid-1800s, when the main industrial applications were railcar axles and steam engines for railways and boats [BAT 10]. The slow rotation of a steam engine was about 50 cycles per minute, more or less. Thus, the fatigue limit was defined by Wöhler to be between 10^6 and 10^7 cycles, but it seems that the quasi-hyperbolic stress number of cycle (SN) curve was suggested by Basquin [BAS 10]. Today, the fatigue life of a high-speed train ranges in the gigacycle, 10^9, regime and for an aircraft turbine it is of the order of 10^{10} cycles, according to the rotation speed of several thousand turns per minute.

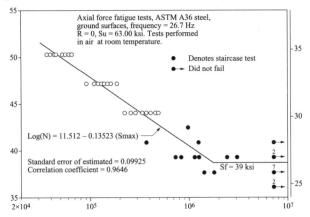

Figure 1.1. *International standard for SN curve and fatigue limit*

The fatigue curve or SN curve is usually defined in reference to carbon steel. The SN curve is generally limited to 10^7 cycles and it is acknowledged, according to the standard, that a horizontal asymptote allows us to determine a fatigue limit value for an alternating stress between 10^6 and 10^7 cycles. Beyond 10^7 cycles, the standard considers that the fatigue life is infinite. For other alloys, it is assumed that the asymptote of the SN curve is not horizontal.

A few results for fatigue limit based on 10^9 cycles can be found in the literature [BAT 10]. Using standard practice, the shape of the SN curve beyond 10^7 cycles is predicted using the probabilistic method, and this is also true for the fatigue limit. In principle, the fatigue limit is given for a number of cycles to failure (Figure 1.1). Using, for example, the staircase method, the fatigue limit is given by the average alternating stress σ_D and the probability of fracture is given by the standard deviation of the scatter (s). The classical way to determine the infinite fatigue life is to use a Gaussian function. Roughly speaking, it is said that σ_D minus $3s$ gives a probability of fracture close to zero. Assuming s is equal to 10 MPa, the true infinite fatigue limit should be $\sigma_D - 30$ MPa. However, experiments show that between σ_D for 10^6 and σ_D for 10^9, the difference is greater than 30 MPa for many alloys.

The so-called standard deviation (SD) approach to the average fatigue limit is certainly not the best way to reduce the risk of rupture in fatigue. When one is conscious that it is the last resort, only experience can remove this ambiguity by appealing to some tests of accelerated fatigue. Today some piezoelectric fatigue machines are very reliable, capable of producing 10^{10} cycles in less than one week, whereas the conventional systems require more than 3 years of tests for only one sample.

To summarize the present situation, it is acknowledged that the concept of a fatigue limit is bound to the hypothesis of the existence of a horizontal asymptote on the SN curve between 10^6 and 10^7 cycles (Figure 1.1). Thus, a sample that reaches 10^7 cycles and is not broken is considered to have an infinite life; that is, in fact, a convenient and economical approximation but not a rigorous approach. It is important to understand that if the staircase method is popular today to determine the fatigue limit, this is because of the convenience of this approximation. A fatigue limit determined by this method to 10^7 cycles requires 30 h of tests to get only one sample with a machine working at 100 Hz. To reach 10^8 cycles, 300 h of tests would be required, which is expensive. Using a 20 kHz piezoelectric fatigue machine, it takes around 14 h to obtain 10^9 cycles, 6 days for 10^{10} cycles and 58 days for 10^{11} cycles. The basic design of the piezoelectric fatigue machine is the same at 30 kHz as a 20 kHz piezoelectric fatigue machine, where the vibration of the specimen is induced by a piezoceramic converter, which generates acoustic waves in the specimen through a power concentrator (horn) in order to obtain desired displacement and an amplification of the stress [WU 93]. The resonant specimen dimension and stress concentration factor were calculated by the Finite Element Method (FEM) subject to 20 and 30 kHz [WU 93]. Such computer-controlled piezoelectric fatigue machines are able to work in tension-compression, tension-tension-tension, bending and torsion loading (Figure 1.2). It is of importance to note that the temperature of the specimen and the amplitude of the stress must stay constant during a standard test at 20 kHz to keep the comparison with low-frequency testing. A complete description of the procedure is given in [BAT 04].

Figure 1.2. *Experimental system for ultrasonic fatigue at 20 kHz*

1.2. Absence of an asymptote on the SN curve

Generally speaking, it is assumed that the steel SN curves are different from the others. To get an overview of the gigacycle behavior, many alloys, including steel, are considered in this chapter. For results of fatigue SN curves based on 10^9 cycles, a few results are available in the literature. Many of those results come from our laboratory [BAT 04]. The other results come from Japanese researchers such as Naito [NAI 84], Kanazawa [NIS 97], Murakami [MUR 99] and Sakai [SAK 07]. They are limited to 10^8 cycles. Also, some SN curves for light alloys come from the laboratory of S. Stanzl-Tschegg and H.R. Mayer [STA 99]. They are limited to 10^9 cycles.

Safe-life design based on the infinite life criteria was initially developed from the Wöhler approach, which is the stress-life or SN curve related to the asymptotic behavior of steel. Some materials display a fatigue limit or an "endurance" limit at a high number of cycles (typically >10^6). Most other materials do not exhibit this response; instead, they display a continuously decreasing stress-life response, even at a large number of cycles (10^6–10^9), which is more correctly described by fatigue strength at a given number of cycles.

The actual shape of the SN curve between 10^6 and 10^{10} cycles is a better way to help the prediction of risk in fatigue cracking (Figure 1.3).

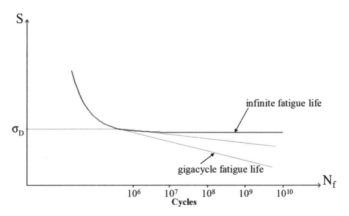

Figure 1.3. *The concept of gigacycle fatigue SN curve*

Since Wöhler, the standard has been to represent the SN curve by a hyperbole more or less modified as indicated below.

Hyperbole $\text{Ln } Nf = \text{Log } a - \text{Ln } \sigma a$, while other methods may be listed as:

– Wöhler $\text{Ln } Nf = a - b \, \sigma a$;

– Basquin $\text{Ln } Nf = a - b \, \text{Ln } \sigma a$;

– Stromeyer $\text{Ln } Nf = a - b \, \text{Ln } (\sigma a - c)$;

Only the exploration of the life range between 10^6 and 10^{10} cycles will create a safer approach to modeling.

1.3. Initiation and propagation

It is of great importance to understand and predict a fatigue life in terms of crack initiation and small crack propagation. It has been generally accepted that at high stress levels, fatigue life is determined primarily by crack growth, while at low stress levels, most of the life is consumed by the process of crack initiation. In low cycle fatigue, it is generally understood that about 50% of the life is devoted to initiation of the micro crack. But many authors demonstrated that the portion of life attributed to crack nucleation is the upper 90% in the high cycle regime (10^6–10^7 cycles) for steel, aluminum, titanium and nickel alloys. In the case for which the crack nucleates from a defect, such as an inclusion or pore, it is said that a relation must exist between the fatigue limit and the crack growth threshold.

However, the relation between the crack growth and initiation is not obvious for many reasons. First, it is not certain that a fatigue crack grows immediately at the first cycle from a sharp defect. Second, when a defect is small, a short crack does not grow as a long crack. In particular, the effect of ratio R or the closure effect depends on the crack length. Thus, the relationship between ΔK_{th} and σ_D is still to be discussed (BAT 00).

Another important aspect is the concept of infinite fatigue life. It is understood that below ΔK_{th} and below σ_D the fatigue life is infinite. In fact, the fatigue limit σ_D is usually determined for Nf = 10^7 cycles. Since the fatigue failure can appear up to or beyond 10^9 cycles, the fatigue strength difference at 10^7 and 10^9 cycles could be more than 100 MPa. This means that the relationship between σ_D and ΔK_{th} must be established in the gigacycle regime if any relation exists.

1.4. Fatigue limit or fatigue strength

How can we model the fatigue limit or the gigacycle fatigue strength of industrial alloys?

The procedure is given below.

First, a new SN curve must be determined up to 10^{10} cycles, which is, in fact, more than the fatigue life of most technological machines.

Second, new fatigue strength at 10^9 cycles has to be predicted using regular statistical method.

In more detail, the prediction of gigacycle fatigue is based on two different mechanisms:

– Initiation is related to flaws (inclusions, defects, pores): prediction is derived from stress concentration, fracture mechanics or short crack approaches.

– Initiation is not related to defect: in this case, microstructure is a key parameter, such as grain size, interface, load transfer and microplasticity.

Thus, the discussion of gigacycle fatigue prediction is split into two parts. The first part is devoted to alloys with flaws.

1.5. SN curves up to 10^9 cycles

In specialized literature, few results were given on this topic until "Euromech 382" was held in Paris in June 1998. Typical gigacycle SN curves are given below as examples.

The experimental results (Figures 1.4–1.6) show that specimens can fail up to 10^9 and beyond. It means the SN curve is not an asymptotic curve. Thus, the concept of infinite life fatigue is not correct and the definition of a fatigue limit at 10^6–10^7 cycles is not conservative [BAT 99]. In Figures 1.4–1.6, it is shown that fatigue failure can occur after 10^{10} cycles in cast aluminum, in SG cast iron and in bearing steel. Depending on the alloy, the difference between the fatigue strength at 10^6 and 10^9 cycles can range from 50 to 200 MPa. From the practical point of view, the gigacycle fatigue strength becomes the most realistic property for predicting very long life [BAT 10].

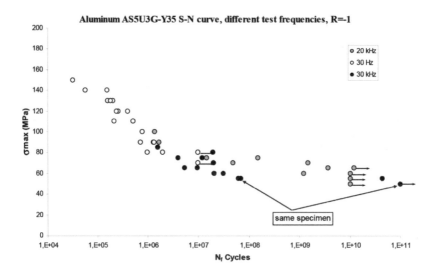

Figure 1.4. *Gigacycle SN curve for a cast aluminum alloy*

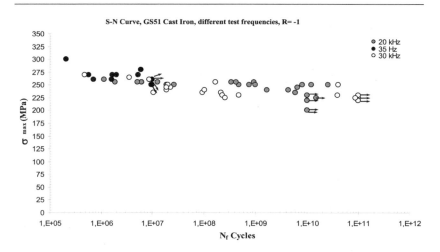

Figure 1.5. *Gigacycle SN curve for an SG cast iron R = −1*

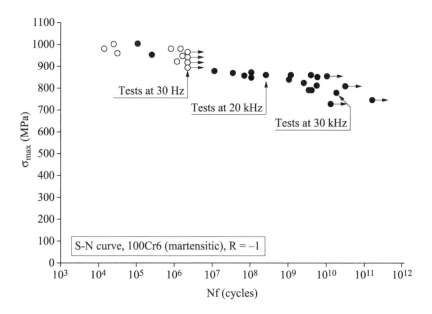

Figure 1.6. *Gigacycle fatigue, 51200 bearing steel. R= −1*

There are only two ways to obtain fatigue data in the gigacycle regime: the rotating bending test (limited to 10^8 cycles) or the high-frequency test. On the contrary, all the fatigue results in the megacycle regime are obtained with low-frequency fatigue machines (1 to 100 Hz). Thus, several questions arise:

– What is the effect of frequency at 20 kHz?

– What is the effect of temperature increase at 20 kHz?

– What is the metal response at high frequency?

– What is the effect of bending, if any, in rotating bending, well known to emphasize the surface plasticity?

These questions need to be clarified and perhaps a new standard needs to be agreed to determine the complete SN curve.

Very often, there is a subsurface initiation site transition beyond 10^7 cycles when inclusion or microstructural defects are operating. Before 10^6 cycles, the initiation of the crack occurs at the surface. But this mechanism is not unique. It means that the surface damage is not always the key mechanism as it is often said, but, in case of no critical defect at this interior, the initiation can appear at the surface in the gigacycle fatigue regime, for example, in pure copper or in pure iron.

1.6. Deterministic prediction of the gigacycle fatigue strength

From a practical point of view, it can be said that the infinite fatigue life for many components is 10^9 or 10^{10} cycles. Thus, it seems that the gigacycle fatigue strength is a crucial property of metal. These data can be determined either by using a statistical approach (staircase) or by using a deterministic relation between stress and defect.

Few models are able to predict the effect of non-metallic inclusions on fatigue strength. This may be because adequate reliable

quantitative data on non-metallic inclusions are hard to obtain. Murakami *et al.* [MUR 99] have investigated the effects of defects, inclusions and flaws on fatigue strength of high-strength steel and expressed the fatigue limit as functions of Vickers hardness (HV) (Kgf/mm^2) and the square root of the projection area of an inclusion or small defect: √area (μm). The fatigue limit prediction equation proposed by Murakami is as follows:

$$\sigma_w = \frac{C(HV+120)}{(\sqrt{area})1/6}\left[\frac{(1-R)}{2}\right]\alpha$$

C = 1.45 for a surface inclusion or defect;

C = 1.56 for interior inclusion or defect;

$\alpha = 0.226 + HV \times 10^{-4}$.

The model does not specify the number of cycles for which the stress σ_w is represented.

According to experimental data, a modified empirical equation, based on Murakami's model, was proposed to estimate the gigacycle fatigue initiation from inclusion and small defects. This model is especially accurate for high-strength steel [WAN 99]. Murakami's parametrical model is an interesting and practical approximation for engineers. Nevertheless, we can see in the results that the error between experience and prediction can sometimes reach up to 26% and it is often about 10%. One difficulty is the estimation of the size of the defect using the square root of the projected area. Table 1.1 shows a comparison between the fatigue strength predicted by the Murakami equation and the experimental results (staircase method) in the gigacycle regime.

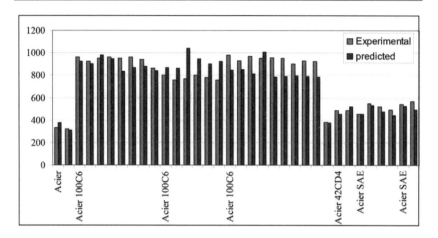

Table 1.1. *Comparison between predicted and experimental fatigue strength at 10^9 cycles*

1.7. Gigacycle fatigue of alloys without flaws

What happens in alloys without inclusion in the gigacycle fatigue regime?

It is well known that in titanium alloys there is not any inclusion or porosity. In this condition, initiation of fatigue crack cannot nucleate from defects; but in titanium alloys, it is found that crack initiation can occur up to 10^9 cycles, despite there not being any inclusion or pore. Figure 1.7 presents SN curves depending on the thermal processing of a Ti 6246 alloy.

It is emphasized that the gigacycle fatigue regime is not always correlated with defects such as inclusions or pore. For Ti 6246, the gigacycle fatigue strength is associated with the transformed amount and secondary alpha volume fraction. Internal fatigue initiation with quasi-cleavage facets in primary alpha phase has been shown [JAG 99, TAO 96].

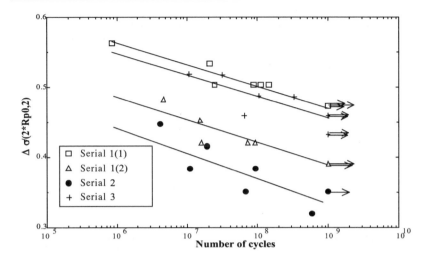

Figure 1.7. *SN curves for Ti alloy with several microstructures [JAG 99]*

1.8. Initiation mechanisms at 10^9 cycles

What remains is to specify how and why some fatigue cracks can initiate inside the metal in gigacycle fatigue. The explanation of the phenomenon is not obvious. It seems that the cycle plastic deformation in plane stress condition becomes very small in the gigacycle regime. In this case, internal defects or large grain size play a role, in competition with the surface damage. It also means that the effect of the environment is quite different in the gigacycle regime since the initiation of short cracks is inside the specimen. Thus, the surface of technological alloys plays a minor role if it is smooth. The effect of plane stress plasticity is evanescent compared to microplasticity due to defects or microstructure misfits. It means that internal initiation is correlated with stress concentration or load transfer.

1.9. Conclusion

It is shown that beyond 10^7 cycles, fatigue rupture can still occur in a large number of alloys. In some cases, the difference of fatigue resistance can decrease by 100, even 200 MPa, between 10^6 and 10^9 cycles to failure. According to our observations, the concept of an infinite fatigue life on an asymptotic SN curve is not correct. Under these conditions, a fatigue limit defined with a statistical analysis between 10^6 and 10^7 cycles cannot guarantee an infinite fatigue life.

Assuming that the fatigue life of engineering components and structures can range above 10^8 cycles, it is very important to determine safe fatigue strength for 10^9 cycles for predicting very long fatigue life of modern components. From a practical point of view, the only way is to use a piezoelectric fatigue machine.

1.10. Bibliography

[BAS 10] BASQUIN O.H., "The exponential law of endurance tests", *Proceedings of the American Society for Testing and Material*, vol. 10, pp. 625–630, 1910.

[BAT 99] BATHIAS C., "There is no infinite fatigue life in metallic materials", *Fatigue and Fracture of Engineering Materials and Structures,* vol. 22, pp. 559–565, 1999.

[BAT 00] BATHIAS C., "Relation between endurance limits and thresholds in the field of gigacycle fatigue", *American Society for Tetsing and materials/Special Technical Publication*, vol. 1372, pp. 135–154, 2000.

[BAT 04] BATHIAS C., PARIS P.C., *Gigacycle Fatigue in Mechanical Practice*, Section 7, Marcel Dekker, 2004.

[BAT 10] BATHIAS C., PINEAU A., *Fatigue of Materials and Structures*, ISTE, London and John Wiley and Sons, New York, 2010.

[JAG 99] JAGO G., BECHET J., "Relation between microstructure and properties in Ti alloys", *Fatigue and Fracture of Engineering Materials and Structures*, vol. 22, pp. 647–655, 1999.

[KAN 97] KANAZAWA K., NISHIJIMA S., "Fatigue fracture of low alloy steel at ultra-high cycle regime under elevated temperature conditions", *Journal of the Society for Materials Science,* vol. 46, no. 12, pp. 1396–1401, 1997.

[MUR 99] MURAKAMI Y., NAMOTO T., UEDA T., "Factors influencing the mechanism of superlong fatigue failure in steels", *Fatigue and Fracture of Engineering Materials and Structures*, vol. 22, pp. 581–590, 1999.

[NAI 84] NAITO T., UEDA H., KIHUSHI M., "Fatigue behavior of carburized steel with internal oxides and nonmartensitic microstructure near the surface," *Metall. Trans. A*, vol. 15, no. 7, pp. 1431–1436, 1984.

[SAK 07] SAKAI T., "Review and prospects for current studies on very high cycle fatigue of metallic materials for machine structural use", *4th International Conference on Very High Cycle Fatigue (VHCF-4)*, TMS (The Minerals, Metals & Materials Society), 2007.

[STA 99] STANZL-TSCHEGG S., "Fracture mechanisms and fracture mechanics at ultrasonic frequency", *Fatigue and Fracture of Engineering Materials and Structures*, vol. 22, pp. 567–579, 1999.

[TAO 96] TAO H., Ultrasonic fatigue at cryogenic temperature, PhD Thesis, CNAM, Paris, 1996.

[WAN 99] WANG Q.Y., BERARD J.Y., BATHIAS C., "Gigacycle fatigue of ferrous alloys", *Fatigue and Fracture of Engineering Materials and Structures,* vol. 22, no. 8, pp. 667–672, 1999.

[WU 93] WU T.Y., NI J.G., BATHIAS C., "Automatic in ultrasonic fatigue machine to study low crack growth at room and high temperature", *ASTM STP*, vol. 1231, pp. 598–607, 1993.

2

Plasticity and Initiation in Gigacycle Fatigue

In Chapter 2, it is shown that intiation in the gigacycle fatigue regime is related to microplasticity, stress field, defects, and metallurgical microstructure. The number of cycles for initiation is several orders of magitude higher than the number of cycles of crack growth.Therefore the damage tolerance is difficult to apply in gigacycle fatigue.

2.1. Evolution of the initiation site from LCF to GCF

The initiation of fatigue crack can be considered differently from a physical or mechanical viewpoint. At the microscopic level, Mughrabi [MUG 06] shows that the initiation of fatigue crack in the gigacycle fatigue (GCF) regime of pure metals can be described in terms of a microstructurally irreversible portion of the cumulative cycle strain. It means that there is no basic difference between fatigue mechanisms in low-cycle fatigue (LCF), megacycle fatigue (MCF) and GCF except for the strain localization. For example, in Armco iron, in LCF, three typical mechanisms are observed depending on the number of cycles [BAT 10]. At the beginning of the test, Franck–Read sources are operating, inducing an increase in the dislocation density until the formation of persistent slip band (PSB) as shown in Figures 2.1 and 2.2. In this case, the macroscopic crack is nucleated from the PSB and from the grain boundary. But, when the crack is growing in opening mode, dislocation cells are formed inside the crack tip opening displacement (CTOD) zone, as shown in Figure 2.3. From a mechanical viewpoint, it is said, roughly speaking, that the mechanisms are controlled by plasticity in plane stress during initiation and plasticity in plane strain during propagation.

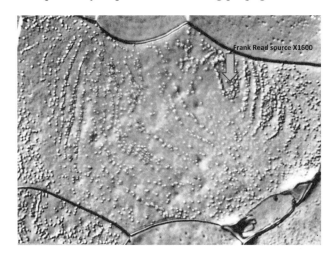

Figure 2.1. *Franck–Read source in Armco iron after few cycles in LCF*

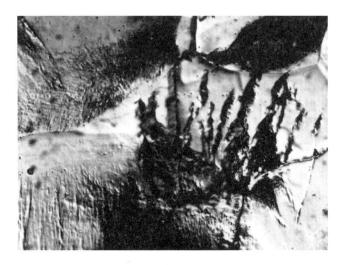

Figure 2.2. *Persistent slip band (PSB), in Armco iron, in LCF*

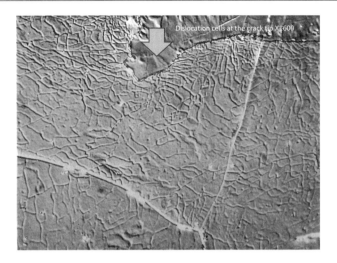

Figure 2.3. *Dislocation cells at the crack tip, in Armco iron, in LCF*

However, specific mechanisms can occur depending on the fatigue life. The fatigue life seems to be a key parameter in correctly determining the fatigue initiation location. In LCF, MCF and GCF, different mechanisms can operate on different scales of plasticity (Figure 2.4).

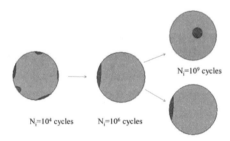

Figure 2.4. *Schematic location of the fatigue crack initiation in LCF, MCF and GCF*

In the LCF regime, the cyclic plastic deformation is critical at the surface but also exists in the bulk of the metal [PIN 10]. Typically, several cracks nucleate from the surface. When the fatigue life is less than 10^5 cycles, general plastic deformation of the specimen bulk governs the initiation. When the fatigue life is between 10^6 and 10^7

cycles, the plastic deformation depends on the plane stress surface effect and the presence of flaws, which explain the critical location of fatigue initiation. Typically, the initiation starts from the surface with only one crack. When Nf approaches 10^9 cycles, the plastic deformation in plane stress conditions vanishes; the macroscopic behavior of the metal is elastic except around flaws, metallurgical defects or inclusions. In a very high-cycle fatigue, the plane stress conditions are not sufficient for a surface plastic deformation. The initiation may be located in an internal zone. When the crack initiation site is in the interior, this leads to the formation of one fish eye on the fracture surface, typical of GCF. In this case, the cyclic plastic deformation is related to the stress concentration around a defect: inclusion, porosity, super grain, etc. Since the probability of occurrence of a flaw is greater within an internal volume than at a surface, the typical initiation in GCF will be most often in the bulk of the metal. However, in pure metals, such as iron or copper, the initiation location is always at the surface because the small inclusions are not critical in low-yield stress behavior, whatever the number of cycles to failure.

According to our own observations and those in the literature, two main factors operate in the gigacycle regime:

– The stress concentration due to metallurgical microstructure misfit becomes an important parameter when the applied load is low. The inclusions inside steels or nickel base alloys and the porosities in cast aluminum or powder metallurgy are among efficient concentrators.

– The stress concentration due to the anisotropy of metals is another parameter. At low deformation, the plasticity can appear only if the grain orientation and the grain size are in agreement with the dislocation sliding. A large grain can become a critical location for initiation in low-carbon steel, austenitic steel, titanium alloy or pure copper.

2.2. Fish-eye growth

2.2.1. *Fracture surface analysis*

When the crack initiation site is in the interior, it leads to the formation of a "fish eye" on the fracture surface, and the origin of the fatigue crack is an inclusion, a large grain (microstructural homogeneity) or a porosity. The crack growth in a fish eye always follows a circular contour whatever the initial shape of the defect. At the macroscopic scale, under the optical microscope (OM) (or naked eye), the fish-eye area appears white, whereas the region outside of the fish eye appears gray, with a dark area in the centre, inside which the crack initiation site is located. Controversy surrounds the origin of this dark area, which some authors have named: "optically dark area (ODA)" [MUR 04] "fine granular area (FGA)" [SAK 07] "granular bright facet (GBF) [SHI 04]. According to Murakami [MUR 02], the mechanism of ODA formation is presumed to be microscale fatigue fracture caused by cyclic stress coupled with internal hydrogen trapped by non-metallic inclusions.

It is presumed that when the size of an ODA [MUR 02] exceeds the critical size for the intrinsic material fatigue strength in the absence of hydrogen, the crack grows without the assistance of hydrogen and cannot become non-propagating. According to Sakai *et al.*, [SAK 07] the mechanism of FGA formation is caused by intensive polygonization induced around the inclusion, followed by microdebondings that can coalesce leading this FGA (Figure 2.5).

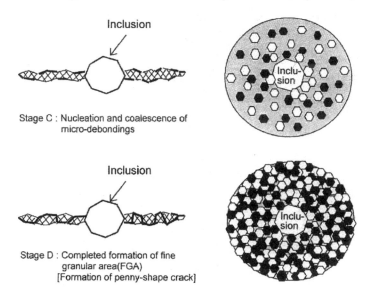

Figure 2.5. *Fine grain area (FGA) model from T. Sakaï*

Using scanning electron microscope (SEM), the fractographic observations (Figure 2.6) do not reveal any dark area around the defect. However, several zones are detected, in agreement with the bases of a mechanical model:

– An FGA due to the initiation mechanisms.

– A penny-shaped zone (short crack growth).Whatever the crack initiation site (spherical or elongated inclusion, supergrain, pore), the fracture surface becomes circular around the initiation site.

– A transition zone with small radial ridges corresponding to the short-to-long crack transition where the fatigue stress intensity factor becomes of the order of the threshold.

– A zone with large radial ridges (long crack growth). In this zone, the fatigue crack propagation produces striations for which the mean distance between striations is a function of ΔK^2, in good agreement with the CTOD model [HUA 11].

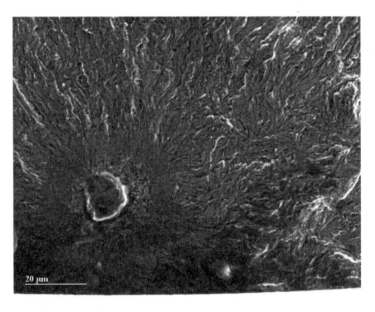

Figure 2.6. *Typical fish-eye initiation with fine grain area, short crack growth and long crack growth in martensitic steel*

2.2.2. *Plasticity in the GCF regime*

In general, the ultrasonic tests are performed with round specimens. Using flat specimens is not so easy as compared to round specimens, and it requires some investigations. From a mechanical viewpoint, it is better to use cylindrical specimens avoiding edge effect.

However, for microscopic observation, a flat specimen is more convenient (Figure 2.7). It must be pointed out that a plane stress field is emphasized in a flat specimen of 1 mm thickness or less. On the contrary, the plane stress effect is limited to the surface in a round specimen. To focus the attention on this effect, some results of tests carried out with round and flat specimens machined in Armco iron and high-strength steel are presented here. A good exemple is to show how the initiation can be nucleated from PSB in iron or from fish eye in martensitic steel, whatever the crucial parameters in the GCF regime.

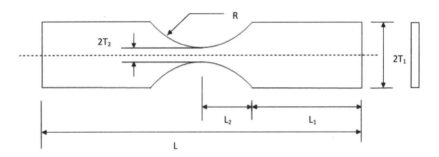

Figure 2.7. *Flat specimen drawing for ultrasonic fatigue*

Working at 20 kHz frequency, a flat specimen will be designed for natural resonance frequency with the same approach for a cylindrical specimen. According to the longitudinal elastic wave equation for a one-dimensional elastic body, and based on the given geometric and the material properties, the resonance length and maximum stress in the middle of the specimen are calculated with the equations given below:

$$L_1 = \frac{1}{k}\arctan\{\frac{1}{k}[\beta\coth(\beta L_2) - \alpha]\}$$

$$\sigma_{max} = E_d A_0 \beta \frac{\cos(kL_1)\exp(\alpha L_2)}{\sinh(\beta L_2)}$$

where

$$C = \sqrt{E_d/\rho}, \quad k = \frac{2\pi f}{C}, \quad \alpha = \frac{1}{2L_2}\ln(\frac{T_2}{T_1}), \quad \beta = \sqrt{a^2 - k^2}$$

where E is the Young modulus and ρ is the density. The first example deals with a single-phase Armco iron for which the ultimate tensile strength (UTS) is 300 MPa and the yield strength is 220 MPa. This is a reference of microplasticity in the gigacycle regime, with a simple microstructure of ferrite grains, without significant inclusion [CHO 13].

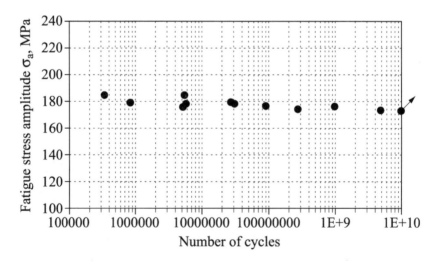

Figure 2.8. *Gigacycle SN curve for Armco iron. Notice a failure after 10^9 cycles [CHO 13]*

A comparison is suggested with very high-strength martensitic steel with the same specimens in the GCF regime (Figures 2.9–2.11).

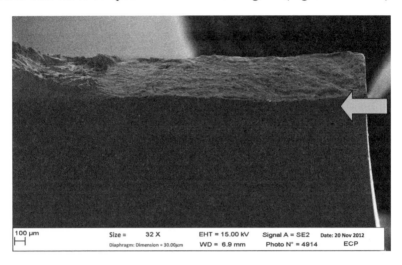

Figure 2.9. *Initiation of a crack in an iron flat specimen. PSB starting at the surface*

The iron is loaded below the yield point at 70, 85 and 120 MPa in push–pull loading at 20 kHz for a failure beyond 10^8 cycles. The fatigue strength is 10% less at 10^9 cycles than at 10^6 cycles. Once again, the concept of a hyperbolic stress number of cycle (SN) curve is not correct even for very low carbon steel. Figures 2.9 and 2.11 show that the initiation occurs at the surface of the flat specimen and round specimen made up of iron.

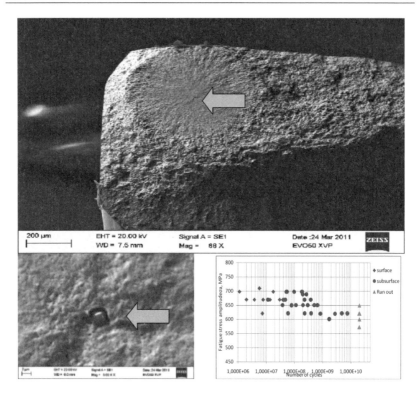

Figure 2.10. *Initiation of a crack in martensitic flat specimen.*
Fish eye at the interior. For a color version of the figure, see
www.iste.co.uk/bathias/Fatigue.zip

Figure 2.11. *Initiation of a crack in round specimen of iron.*
Quasi-PSB starting from the surface

The damage starts in stage 1 along more or less 1 mm depth before growing in stage 2. In this case, there is not any difference between MCF and GCF. During stage 1, it is clear that the microscopic mechanism is related to the formation of quasi-PSBs [MUG 76, LUK 65] (Figure 2.12).

Figure 2.12. *PSB at the surface of an iron flat specimen at 70 Mpa, R = −1 (Chong Wang)*

It seems that a threshold exists, in iron, for the formation of the quasi-PSB. It is found that below 70 Mpa, no PSB, occurs up to 10^9 cycles. However, the number of PSBs increases with the number of cycles and no experiment exists to prove that some PSBs, cannot form at 10^{10} cycles or more. But, no fish eye occurs in these conditions, probably due to the strong effect of the plane stress field even if the surface is electropolished.

It is significant to compare iron and martensitic steel, with the same flat specimen (1 mm) and the same frequency, in push–pull loading, for a fatigue life beyond 10^8 cycles. Figure 2.10 shows that the initiation starts from a fish eye in the martensitic steel and not from the surface at 620 MPa. Why this difference?

The fish eye occurs in the high-strength steel from a small inclusion of oxide, in plane strain conditions, in spite of the small thickness, with a plasticity located inside a plastic zone around an inclusion. In this case, the plane stress effect at the surface is not efficient because of the stress concentration due to a defect in this steel where the UTS is close to 2,000 MPa. There is competition between plane strain plasticity and plane stress plasticity. In this case, the microplasticity is not only governed by the von Mises criteria, but also by the stress concentration effect at the tip of the inclusion. In plane strain plasticity, the PSB or quasi-PSB is not observed in the fish eye. As suggested by T. Sakaï, the mechanisms of plasticity seem relevant for polygonization, cell formation, grain refining or phase transformation around the inclusion.

These results show that the initiation in the GCF is explained by a fish-eye formation except when the effect of plane stress is dominating in a thin sheet or in a bending bar when the yield point of the metal is low. In this condition, the surface is governing the initiation. Otherwise, in most alloys, the initiation, in the gigacycle range, is beneath the surface and always depends on defects: inclusions, pores and large grain. The effect of the surface is less important, for example in the case of high strength steels. According to these observations, more attention must be paid to the microplasticity inside the fish eye in high-strength alloys such as martensitic steels.

The very high fatigue life, called gigacyclic, requires more attention for the choice of the alloys or for the prediction of endurance. An approach-based ΔK and the crack concept are not suitable due to frequent occurrence of an incubation nucleation process [BAT 00].

To solve this problem, it has been suggested to follow the Paris–Hertzberg law for the formation of the so-called fish eye, which is typical of damage in the GCF of metals. In other words, the fatigue strength in the gigacycle regime can be predicted using a Murakami-type formula taking into account the size of the flaw from which the fatigue crack nucleates. The last point is the response of the microstructure around a flaw, in other words a stress concentration, after a great number of cycles applied in the case of ultrasonic fatigue with a high strain rate. The instability of the microstructure is often related to a heating dissipation with an increase in the temperature. The increase in temperature associated with a microstructural transformation seems to explain a frequency effect that does not exist for many stable alloys (see Chapter 3). To predict the fish-eye nucleation and growth, we should refer to the general behavior pattern of the stress concentration around flaws.

2.3. Stresses and crack tip intensity factors around spherical and cylindrical voids and inclusions

In GCF of industrial alloys, crack initiation and growth almost always occurs from internal defects in the materials including voids, second phases and inclusions. Occasionally, a surface defect of hemispherical shape is also encountered. To understand the stresses near these imperfections and the stress intensity factors for cracks initiating from them, some elastic stress formulas will be developed by Paris and co-workers [PAR 11]. For the inclusions, mismatches in elastic properties and sizes will be treated for realistic examination of their effects (Figure 2.13). It is expected that the availability of such formulas may enhance an understanding of GCF initiation and crack growth.

2.3.1. Spherical cavities and inclusions

Under uniaxial stress, σ, the *spherical cavity* will have a stress concentration factor, K_t, which is defined by:

$$\sigma_{\max} = K_t \sigma \qquad [2.1]$$

In this case, the concentration factor is given in standard texts on the theory of elasticity as:

$$K_t = 3/2\left(1 + \frac{2}{7 - 5\nu}\right) \qquad [2.2]$$

where ν is Poisson's ratio.

However, for *triaxial tension* $\bar{\sigma}$, the stress concentration factor is given by:

$$K_t = \frac{3}{2} \qquad [2.3]$$

Moreover, if instead of an internal spherical cavity, the *hemispherical surface cavity* is the case of significance, the increase in the stress concentration factor is less than 2% for the uniaxial case or:

$$K_t = 1.522\left(1 + \frac{2}{7 - 5v}\right) \qquad [2.4]$$

The stress outside the *spherical cavity under uniaxial loading* is given at a radial distance r compared to the radius of the sphere R by:

$$\sigma_{axial} = \left[1 + \frac{4 - 5v}{14 - 10v}\frac{R^3}{r^3} + \frac{9}{14 - 10v}\frac{R^5}{r^5}\right]\sigma \qquad [2.5]$$

Further, for the *spherical cavity under triaxial loading* $\bar{\sigma}$ with *internal pressure p*, the result is:

$$\sigma_\theta = \bar{\sigma} + \frac{(\bar{\sigma} + p)R^3}{2r^3} \qquad [2.6]$$

Both of these formulas give the stresses on a prospective crack plane extending outward from the cavity. For the latter case, in equation [2.6], the radial displacement from the surface of the cavity outward is:

$$u_{r-cavity} = \frac{(1 + v)(\bar{\sigma} + p)}{2E}\frac{R^3}{r^2} + \frac{(1 - 2v)\bar{\sigma}r}{E} \qquad [2.7]$$

where E is the elastic modulus.

2.3.2. *Spherical inclusion*

For the spherical inclusion with external pressure p, the radial displacement of the surface is:

$$u_{r-inclusion} = -\frac{(1-2v')\,pR}{E'} \qquad\qquad [2.8]$$

where E' and v' are the elastic constants of the inclusion.

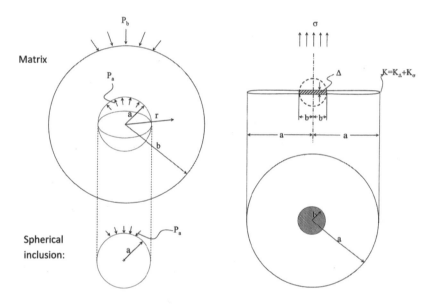

Figure 2.13. *Spherical bonded inclusion and mismatch inclusion (PC Paris)*

2.3.3. *Mismatched inclusion larger than the spherical cavity it occupies*

If the spherical inclusion is larger than the spherical cavity it occupies, then there will be a contact pressure p, which will also depend on the external hydrostatic tension, $\bar{\sigma}$. The mismatch will be that of an inclusion that is larger by a radial amount, Δ. Then, the compatibility of the radial displacements between the cavity and inclusion can be expressed as:

$$u_{r-cavity} = u_{r-inclusion} + \Delta \quad (\text{at } r = R) \qquad [2.9]$$

Combining equations [2.7]–[2.9] leads to an additional stress outside the cavity as:

$$\sigma_\theta(r = R) = \frac{1}{\dfrac{1+v}{2E} + \dfrac{1-2v'}{E'}}\left(\frac{\Delta}{2R}\right) \qquad [2.10]$$

which should be added to the previous stress result equation [2.6] with $r = R$ and $p = 0$. Under such circumstances, the pressure p generated between the inclusion and the void is:

$$p = \frac{\dfrac{\Delta}{R} - \dfrac{3}{2}\dfrac{(1-v)}{E}\,\bar{\sigma}}{\dfrac{3}{2}\dfrac{(1-v')}{E'}} \quad (\text{compression}) \qquad [2.11]$$

Now, if there is no mismatch and the inclusion is bonded to the external body, then the stress concentration is:

$$K_t = \frac{3}{2}\left\{1 - \frac{\dfrac{1-v}{2E}}{\dfrac{1-2v'}{E'} + \dfrac{1+v}{2E}}\right\} \qquad [2.12]$$

When, $E' = 0$, that $K_t = 3/2$; for $E' = E, v' = v$, that $K_t = 1$; and for $E' = \infty$, that $K_t = \dfrac{3v}{1+v}$, which is as expected. With this in mind, the stress in the body next to the inclusion is:

$$\sigma_{\theta-\text{max}}\left(r=R^{+}\right)=K_{t}\overline{\sigma}+\cfrac{1}{\cfrac{1+v}{2E}+\cfrac{1-2v'}{E'}}\left(\cfrac{\Delta}{2R}\right) \qquad [2.13]$$

This form accommodates a bonded inclusion of differing elastic properties with a mismatch in its size compared to the void in the main body. As a first approximation, it is suggested here for the case of uniaxial stress applied to the body that the K_t can be increased by the (3/2) factor in equation [2.2].

2.3.4. *Cylindrical cavities and inclusions*

For cylindrical cavities and inclusions, the case may be plane stress or plane strain depending on constraint conditions. For this reason, it is convenient to use modified elastic constants, G, the shear modulus and β, which depends on Poisson's ratio but changes with the constraint. They are defined as:

$$G=\frac{E}{2\left(1+v\right)} \quad \text{and} \quad \beta=1-v \quad \text{for plane strain and} \quad \beta=\frac{1}{1+v} \quad \text{for}$$

plane stress. For the inclusion, they will be written with a prime. For plane stress, it is assumed that the cylindrical void and the cylindrical inclusion are smooth (frictionless) and unbonded.

For both constraint cases, the biaxial (or triaxial) exterior applied stress is taken as $\overline{\sigma}$, and the contact pressure between the void and cylinder is p.

The classical equations for the stresses in the outer body are:

$$\sigma_{r}(r\geq R)=\overline{\sigma}-\frac{\left(p+\overline{\sigma}\right)R^{2}}{r^{2}} \qquad [2.14]$$

$$\text{and } \sigma_{\theta}(r\geq R)=\overline{\sigma}+\frac{\left(p+\overline{\sigma}\right)R^{2}}{r^{2}} \qquad [2.15]$$

which lead to the radial displacement given by:

$$2Gu_r(r \geq R) = (2\beta - 1)\bar{\sigma}r + \frac{(p + \bar{\sigma})R^2}{r}$$ [2.16]

when the inclusion radius is larger than the void radius. It leads to:

$$u_{r-inclusion}(r = R) + \Delta = u_{r-void}(r = R)$$ [2.17]

which leads to:

$$p = \frac{2}{\frac{1}{G} + \frac{2\beta' - 1}{G'}}\left[\frac{\Delta}{R} - \frac{\beta\bar{\sigma}}{G}\right]$$ [2.18]

This can be used to get the maximum stress in the body, which is:

$$\sigma_{\theta-max}(r = R^+) = p + 2\bar{\sigma}$$ [2.19]

as long as contact is not lost, $p \geq 0$.

Further, the 2 is changed to 3 for uniaxial exterior applied stresses, $\bar{\sigma}$, as a first approximation for that case of stressing. Stress intensity factors at crack tips grow from voids and inclusions, some with misfit sizes. To give estimates of the crack tip stress intensity factors for cracks starting at the voids or inclusions and growing to failure, asymptotic approximation methods will be adopted. It consists of exactly fitting the exact stress intensities for small cracks sizes and also for large crack sizes and then fitting a smooth cubic curve between these exact solutions. For a typical exemple, the results of such a method are presented for a hemispherical void on the surface of a solid, such as might occur due to a corrosion pit.

2.3.5. *Cracking from a hemispherical surface void*

For a spherical radius R and a crack of depth a measured from the surface of the hemisphere, the asymptotic approximation of the crack tip stress intensity factor is (making use of the factor 1.015 for the half-plane effect):

$$K = 1.015\sigma\sqrt{\pi a} \cdot F(x,v) \text{ where } x = \frac{a}{R} \text{ for } 0 \le \frac{a}{R} \le \infty \quad [2.20]$$

where

$$F(x,v) = A(v) + B(v)\frac{x}{1+x} + C(v)\left(\frac{x}{1+x}\right)^2 + D(v)\left(\frac{x}{1+x}\right)^3 \quad [2.21]$$

Further, the coefficients A, B, C and D are found to be:

$$A(v) = 1.683 + \frac{3.366}{7-5v} \qquad B(v) = -1.025 - \frac{12.3}{7-5v}$$

$$C(v) = -1.089 + \frac{14.5}{7-5v} \qquad D(v) = 1.068 - \frac{5.568}{7-5v} \quad [2.22]$$

These coefficients are for both uniaxial stress and biaxial stress, σ, applied parallel to the surface from which the pit emanates. However, the factor 1.015 in equation [2.20] is, for the deepest part of the crack front, away from the surface forming the hemisphere.

With an angle θ measured from a line perpendicular to that surface, the factor 1.015 may be replaced by:

$$f(\theta) = 1.210 - 0.195\sqrt{\cos\theta} \qquad \textbf{for } \left(-80° \le \theta \le +80°\right) \quad [2.23]$$

in order to get the stress intensity factor, K, along the crack front.

Adjustments may be made in these values for K due to the imperfection in the hemispherical shape or unequal depth of the crack, a, around the hemisphere as suggested in Appendix I of [TAD 00]. The stress intensity factor for a crack growing from a misfit size spherical inclusion or a void.

The crack size, a, measured from the surface of the spherical inclusion to the surrounding body will be represented here by $\bar{a}$ measured from the center of the inclusion or void. This is expressed by $\bar{a} = R + a$, where R is the radius of the sphere. This assumes no difference in material properties.

For *very large cracks* ringing the sphere, $\dfrac{\bar{a}}{R} >> 1$, the mismatched spherical inclusion is larger by controlling Δ as the size of the mismatch. For a large ring crack, the load, P, imposed on the exterior body is:

$$P = \frac{2ER\Delta}{1-v^2} \tag{2.24}$$

The stress intensity factor for the opposing concentrated loads, P, in the center of a circular crack is:

$$K_\Delta = \frac{P}{(\pi\bar{a})^{\frac{3}{2}}} = \frac{2ER\Delta}{(1-v^2)(\pi\bar{a})^{\frac{3}{2}}} \tag{2.25}$$

The additional stress intensity factor due to uniform normal stress, σ, perpendicular to the crack (any additional normal stresses parallel to the crack have no effect) is:

$$K_\sigma = \frac{2}{\pi}\sigma\sqrt{\pi\bar{a}}\left(1 - \frac{4R}{\pi^2\bar{a}}\right) \tag{2.26}$$

where the sum of these variables gives the total stress intensity factor for large cracks:

$$K_{total} = K_\Delta + K_\sigma \tag{2.27}$$

On the other hand, *for small cracks* emanating into the surrounding body from the inclusion, $\dfrac{\bar{a}}{R} = \dfrac{R+a}{R} \cong 1^+$, the results are:

$$K_\Delta = \frac{E\Delta}{2\left(1-v^2\right)\sqrt{\pi\left(\dfrac{\bar{a}-R}{2}\right)}}, \qquad K_\sigma = \sigma\sqrt{\pi\frac{\left(\bar{a}-R\right)}{2}}$$

$$K_{total} = K_\Delta + K_\sigma \tag{2.28}$$

From the first expression in equations [2.28] and [2.25], the final asymptotic approximation is:

$$K_\Delta = \frac{E\Delta}{\left(1-v^2\right)\pi\sqrt{\pi\bar{a}}}\left(\frac{\dfrac{R}{\bar{a}}}{\sqrt{1-\dfrac{R}{\bar{a}}}}\right)F_\Delta\left(\frac{R}{\bar{a}}\right) \tag{2.29}$$

where:

$$F_\Delta\left(\frac{R}{\bar{a}}\right) = 1 - 0.5\left(\frac{R}{\bar{a}}\right) + 2.44\left(\frac{R}{\bar{a}}\right)^2 - 1.83\left(\frac{R}{\bar{a}}\right)^3 \tag{2.30}$$

Further, for the asymptotic relationship for σ, the result is:

$$K_\sigma = \sigma\sqrt{\pi\bar{a}}\left(\sqrt{\frac{1-\dfrac{R}{\bar{a}}}{2}}\right)F_\sigma\left(\frac{R}{\bar{a}}\right) \tag{2.31}$$

where:

$$F_\sigma\left(\frac{R}{\bar{a}}\right) = 0.900 + 0.085\left(\frac{R}{\bar{a}}\right) + 0.015\left(\frac{R}{\bar{a}}\right)^2 \tag{2.32}$$

Again, the final approximation for $1 \leq \dfrac{\bar{a}}{R} \leq \infty$ combines equation [2.29] with [2.32] to give:

$$K_{total} = K_\Delta + K_\sigma \qquad [2.33]$$

2.3.6. Crack tip stress intensity factors for cylindrical inclusions with misfit in both size and material properties

For small cracks in the exterior body of length a as compared to radius R of the cylindrical inclusion, or $\dfrac{a}{R} \ll 1$, the stresses at the initial crack site, σ_0, and its reduction at the crack tip, σ_1, as caused by the gradient away from the inclusion are:

$$\sigma_0 = \sigma_\theta (r = R) = 2\bar{\sigma} + p \qquad [2.34]$$

where $\bar{\sigma}$ is the externally applied biaxial stress and p is the contact pressure.

Then:

$$\sigma_1 = \frac{d\sigma_\theta (r=R)}{dr} a = 2\bar{\sigma}\left(1 - 2\beta \frac{G_{eff}}{G}\right)\frac{a}{R} \qquad [2.35]$$

where:

$$\frac{1}{G_{eff}} = \frac{1}{G} + \frac{1}{G'}(2\beta' - 1) \qquad [2.36]$$

Now, *when there is no misfit between the inclusion and exterior body,* and bonded together for plane strain or smooth frictionless contact for plane stress between them, the stress intensity factor is of the form:

$$K_\sigma = \bar{\sigma}\sqrt{\pi a}F_\sigma\left(\frac{a}{R}, \gamma, \frac{G_{eff}}{G}\right) \qquad [2.37]$$

where:

$$F_\sigma\left(\frac{a}{R}, \gamma, \frac{G_{eff}}{G}\right) = 2\left[\left(1+0.122\frac{1}{1+\gamma}\right)\left(1-\beta\frac{G_{eff}}{G}\right) - \frac{2}{\pi}\left(1+0.073\frac{1}{1+\gamma}\right)\left(1-2\beta\frac{G_{eff}}{G}\right)\frac{a}{R}\right]$$

$$[2.38]$$

where: $\gamma = \dfrac{G'}{G}$ and applies for $\gamma = \dfrac{G'}{G}$

Now, *for large cracks*, the values of F_σ asymptotically approach constants that are:

$F_\sigma(\ ) = 1$ (for cracks on both sides of the inclusion)

$F_\sigma(\ ) = \dfrac{1}{\sqrt{2}}$ (for a crack on one side of the inclusion)

Further, for the full range of crack sizes, $0 < \dfrac{a}{R} < \infty$, the cubic asymptotic fit of the end curves will be applied. It is:

$$F_\sigma\left(x, \gamma, \gamma_{eff}\right) = A + B\left(\frac{x}{1+x}\right) + C\left(\frac{x}{1+x}\right)^2 + D\left(\frac{x}{1+x}\right)^3 \quad [2.39]$$

where:

$$A\left(\gamma, \gamma_{eff}\right) = 2\left(1+0.122\frac{1}{1+\gamma}\right)\left(1-\beta\gamma_{eff}\right)$$

$$B\left(\gamma, \gamma_{eff}\right) = -\frac{\pi}{4}\left(1+0.073\frac{1}{1+\gamma}\right)\left(1-2\beta\gamma_{eff}\right) \quad [2.40]$$

For *cracks on both sides* of the inclusion:

$$C\left(\gamma, \gamma_{eff}\right) = 3 - 3A - 2B$$
$$D\left(\gamma, \gamma_{eff}\right) = -2 + 2A + B$$

$$[2.41]$$

where:

$$x = \frac{a}{R}, \quad \gamma = \frac{G'}{G}, \quad \gamma_{eff} = \frac{G_{eff}}{G}$$

However, *for a crack on one side of the inclusion*, equation [2.41] should be replaced by:

$$C = 2.121 - 3A - 2B$$
$$D = -2 + 2A + B$$

[2.42]

This completes the discussion for the externally applied stresses with differing elastic properties of the inclusion.

Next, the case of a radial difference, Δ, interference between the inclusion and its nest in the exterior body will be considered. For small cracks, $\frac{a}{R} \ll 1$, on one side or both sides of the inclusion, the result is:

$$K_\Delta = 2G_{eff} \frac{\Delta}{R} \sqrt{\pi a} F_\Delta\left(\frac{a}{R}, \gamma\right)$$

[2.43]

where:

$$F_\Delta\left(\frac{a}{R}, \gamma\right) = 1 + 0.122\left(\frac{1}{1+\gamma}\right) - \frac{4}{\pi}\left(1 + 0.073\frac{1}{1+\gamma}\right)\frac{a}{R}$$

[2.44]

For large cracks, $\frac{a}{R} \gg 1$, the result for a crack on *one side of the inclusion* is:

$$K_\Delta = \frac{G_{eff}\Delta}{\beta\sqrt{\pi a}}\sqrt{\frac{1}{\frac{a}{R}\left(2 + \frac{a}{R}\right)}}$$

[2.45]

The asymptotic interpolation between these solutions for large and small cracks on *one side of the inclusion* is:

$$K_\Delta = G_{eff} \frac{\Delta}{R} \sqrt{\pi a} F_\Delta(x, \gamma)$$

[2.46]

where:

$$F_\Delta(x,\gamma) = A + B\left(\frac{x}{1+x}\right) + C\left(\frac{x}{1+x}\right)^2 + D\left(\frac{x}{1+x}\right)^3 \qquad [2.47]$$

with:

$$A(\gamma) = 2\pi f_0 \quad B(\gamma) = -8f_1$$

$$C(\gamma) = -6\pi f_0 + 16 f_1 \quad D(\gamma) = 4\pi f_0 - 8 f_1 \qquad [2.48]$$

and:

$$f_0 = 1 + 0.122\left(\frac{1}{1+\gamma}\right), \quad f_1 = 1 + 0.073\left(\frac{1}{1+\gamma}\right) \qquad [2.49]$$

Finally, these results, from equation [2.37] through equation [2.49], may be combined to obtain the total stress intensity factor from:

$$K_{total} = K_\sigma + K_\Delta \qquad [2.50]$$

The intended application of this work is to analyze the initiation and growth of cracks in gigacycle fatigue. According to this goal, equation [2.50] puts in the maximum stress, σ, to obtain the maximum stress intensity factor. For the range of the stress intensity factor, since K_Δ imposes a constant stress intensity factor, the range should be calculated from K_σ only by:

$$\Delta K_{total} = \Delta K_\sigma \qquad [2.51]$$

Indeed, the K_Δ decreases as the crack size, a, increases, so that its local stress affects initiation of cracking, but only slightly increases the rate of crack growth as the crack becomes large compared to the size of the inclusion. Meanwhile, the load ratio, R, changes due to the reduction in K_{total} as compared to ΔK_{total}.

2.4. Estimation of the fish-eye formation from the Paris–Hertzberg law

Nakasone and Hara [NAK 04] have used FEM in order to simulate the crack growth of internal initiation of failures using the Paris law $da/dN = C (\Delta K)^m$. They have found that crack growth life is approximately 10^5 cycles of total lifetime, but they did not include this fact in their conclusions. They found that the smaller portion of life involving internal crack growth is negligible compared to the much longer crack initiation life. They consider the internal environment in materials to be almost vacuum. They assume that this causes a decrease in the crack growth rate by 10 times in comparison to the crack growth in the air environment.

Similarly, Sugeta *et al.* [SUG 04] have performed a calculation of crack growth life for internal flaws (naming this portion of life, NP) with the same expression used by Nakasone and Hara. They also do not show the significance of NP in total lifetime.

As is reported in the ASTM standard, due to the lack of closure, the growth of small cracks is sometimes significantly faster than long cracks at the same nominal crack driving force. Consequently, in this work, the effect of closure will be explored for a range of transition sizes from small to large cracks, as well as various load ratios, to evaluate these effects on the total life.

Paris *et al.* [PAR 04] have shown a historical review on crack growth and threshold to develop the estimation procedure for crack growth life for internal initiation, called "fish eye". Our modeling is supported by the observations and analysis summarized by [PAR 99], and [HER 93, HER 95, HER 97, HER 12] who predict the threshold corner at:

$$\frac{da}{dN} = b \qquad and \qquad \frac{\Delta K_{eff}}{E\sqrt{b}} = 1$$

where b is Burger's vector and E is elastic modulus. The corresponding crack growth law is used to estimate the life of the small cracks in the fish-eye range, considering that crack closure is minimal for this type of crack:

$$\frac{da}{dN} = b\left(\frac{\Delta K_{eff}}{E\sqrt{b}}\right)^3 \qquad\qquad [2.52]$$

Paris [PAR 99] and Hertzberg [HER 93, HER 95] have shown the effectiveness of this relationship as a predictor of the threshold corner and slope (Figures 2.14 and 2.15).

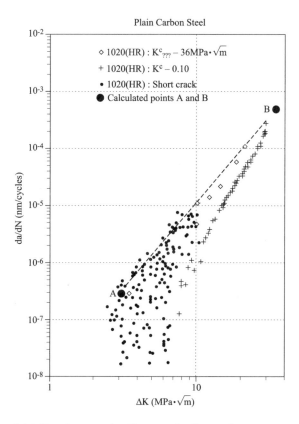

Figure 2.14. *Hertzberg results. Short crack effect on fatigue propagation*

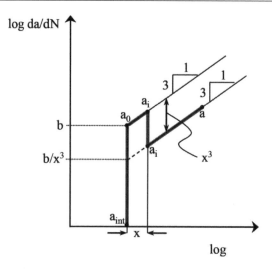

Figure 2.15. *Modeling of the short crack effect for an internal circular crack. For a color version of the figure please see www.iste.co.uk/bathias/Fatigue.zip*

The initiation from an inclusion or other defect itself must be close to the total life, perhaps much more than 99% of the life in many cases. This is made evident by integrating the fatigue crack growth rates for small cracks to estimate the possible extent of crack growth life.

To do this, we should refer to the general behavior pattern of the crack growth rate curve as illustrated by the previous equations. It is noted that small cracks such as those growing from small inclusions do not exhibit crack closure, so these equations in terms of ΔK_{eff} apply fairly well. They form an upper bound on crack growth rates for the small cracks in the "fish-eye" range for which crack closure is minimal (Figures 2.14 and 2.15).

Estimating the life for a crack of this type beginning just above the threshold is then appropriate, considering the growth law as:

$$\frac{da}{dN} = b\left(\frac{\Delta K_{eff}}{E\sqrt{b}}\right)^3$$

[2.52]

where for the circular crack growing from a "fish eye", the stress intensity factor formula is:

$$\Delta K = \frac{2}{\pi} \Delta \sigma \sqrt{\pi a} \qquad [2.53]$$

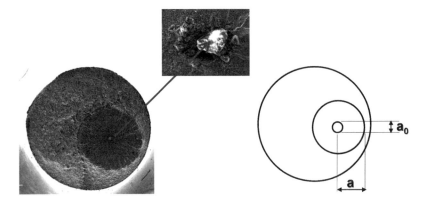

Figure 2.16. *Fish eye in gigacycle fatigue initiation in 4240 steel and modeling. For a color version of the figure, see www.iste.co.uk/bathias/Fatigue.zip*

The integration of the Paris–Hertzberg law without transition, to determine the crack growth life, will begin here with the crack growth rate corner, which we will denote as ΔK_0 corresponding to an initial circular crack of radius a_0 (Figure 2.12). Substituting these into the first formula (equation [2.52]), we obtain:

$$\frac{da}{dN} = b \left(\frac{\Delta K}{E \sqrt{b}} \right)^3 \left(\frac{a}{a_0} \right)^{3/2} = b \left(\frac{a}{a_0} \right)^{3/2}$$

This equation can be integrated from a_0 to a_{final}, which gives:

$$N_f \int_{a_0}^{a_f} \frac{(a_o)^{3/2}}{b} \bullet \frac{da}{(a)^{3/2}} \left[\frac{1}{(a)^{1/2}} - \text{small} \right] \cong \frac{2 a_o}{b}$$

But it can be noted that:

$$1 = \frac{\Delta K_0}{E\sqrt{b}} = \frac{2\Delta\sigma\sqrt{a_o}}{\sqrt{\pi}E\sqrt{b}} \text{ or } a_o = \frac{\pi E^2 b}{4(\Delta\sigma)^2}$$

thus leading to the following formula:

$$N_f = \frac{\pi E^2}{2(\Delta\sigma)^2} \qquad\qquad [2.54]$$

This is the result for the approximate number of cycles from the threshold corner to failure from an initial crack size a_0 as expressed above.

E = dynamic Young's modulus at 20 kHz

$\Delta\sigma$ = experimental nominal stress range

$N_f = N_{fish\text{-}eye} = N_{crack\ growth}$

Estimating the life for an internal "fish-eye" crack beginning just above threshold, it is then appropriate to consider that the growth life ($N_{fish\text{-}eye}$) for an internal "fish-eye" will be defined considering its respective portion of life, from corner threshold (N_{total}) when it is a small crack $\left(N_{a_0 \to a_i}\right)$ and when it changes to a large crack $\left(N_{a_i \to a}\right)$, and its respective crack growth life below threshold (N_{int}):

$$N_{fish\text{-}eye} = N_{total} + N_{int}$$

$$N_{total} = N_{a_0 \to a_i} + N_{a_i \to a}$$

The integration to determine the small crack growth life will begin here with the crack growth rate corner, which we will denote as ΔK_0 corresponding to an initial circular crack of radius a_0 until a circular crack of radius a_i (transition of small crack to large crack). Substituting these into the crack growth law formula, we obtain:

$$\frac{da}{dN} = b\left(\frac{\Delta K_0}{E\sqrt{b}}\right)^3 \left(\frac{a}{a_0}\right)^{3/2} = b\left(\frac{a}{a_0}\right)^{3/2}$$

$$N_{a_0 \to a_i} = \frac{(a_0)^{3/2}}{b} \int_{a_0}^{a_i} \frac{da}{a^{3/2}} = \frac{2(a_0)^{3/2}}{b}\left[\frac{1}{\sqrt{a_0}} - \frac{1}{\sqrt{a_i}}\right] = \frac{2a_0}{b}\left[1 - \sqrt{\frac{a_0}{a_i}}\right]$$

$$N_{a_0 \to a_i} = \frac{\pi E^2}{2(\Delta\sigma)^2}\left[1 - \sqrt{\frac{a_0}{a_i}}\right] \qquad [2.55]$$

2.4.1. "Short crack" number of cycles

The integration to determine the large crack growth life will begin at the transition point "short-to-long crack" with the crack growth rate b/x^3, which we will denote as ΔK_0 corresponding to a circular crack of radius a_i until a final circular crack of radius a. Substituting these into the crack growth law formula, we obtain:

$$\frac{da}{dN} = \frac{b}{x^3}\left(\frac{\Delta K_0}{E\sqrt{b}}\right)^3 \left(\frac{a}{a_0}\right)^{3/2} = \frac{b}{x^3}\left(\frac{a}{a_0}\right)^{3/2}$$

$$N_{a_0 \to a_i} = \frac{x^3 (a_0)^{3/2}}{b} \int_{a_i}^{a} \frac{da}{a^{3/2}} = \frac{2(a_0)^{3/2}}{b}\left[\frac{x^3}{\sqrt{a_i}} - \frac{x^3}{\sqrt{a}}\right] = \frac{2a_0}{b}\left[x^3\sqrt{\frac{a_0}{a_i}} - x^3\sqrt{\frac{a_0}{a}}\right]$$

$$N_{a_0 \to a_i} = \frac{\pi E^2}{2(\Delta\sigma)^2}\left[x^3\sqrt{\frac{a_0}{a_i}} - x^3\sqrt{\frac{a_0}{a}}\right] \qquad [2.56]$$

2.4.2. "Long crack" number of cycles

The integration to determine the below-threshold crack growth life can be estimated by the law used above, from an initial flaw a_{int} to a_0 at the threshold corner; but prior to the corner, threshold should have a very much higher slope α (Figure 2.15).

$$\frac{da}{dN} = b\left(\frac{\Delta K_0}{E\sqrt{b}}\right)^{\alpha}\left(\frac{a}{a_0}\right)^{\alpha/2} = b\left(\frac{a}{a_0}\right)^{\alpha/2}$$

$$N_{int} = \frac{(a_0)^{\alpha/2}}{b}\int_{a_{int}}^{a_0}\frac{da}{a^{\alpha/2}} = \frac{(a_0)^{\alpha/2}}{b\left(\frac{\alpha}{2}-1\right)}\left[\frac{1}{(a_{int})^{\left(\frac{\alpha}{2}-1\right)}} - \frac{1}{(a_0)^{\left(\frac{\alpha}{2}-1\right)}}\right] = \frac{a_0}{b}\frac{1}{\left(\frac{\alpha}{2}-1\right)}\left[\left(\frac{a_0}{a_{int}}\right)^{\left(\frac{\alpha}{2}-1\right)} - 1\right]$$

$$N_{int} = \frac{\pi E^2}{2(\Delta\sigma)^2}\frac{1}{\left(\frac{\alpha}{2}-1\right)}\left[\left(\frac{a_0}{a_{int}}\right)^{\left(\frac{\alpha}{2}-1\right)} - 1\right] \qquad [2.57]$$

2.4.3. "Below threshold" number of cycles

The result to estimate the total crack growth lifetime for an internal failure "fish eye", which is the addition of following lifetimes:

– below threshold, from an initial flaw a_{int} to a_0;

– short crack, from an initial circular crack of size a_0 to a_i;

– long crack, transition from a small crack point to a large crack point, a_i to a.

$$N_{fish-eye} = N_{total} + N_{int}\frac{\pi E^2}{2(\Delta\sigma)^2}\left[1+(x^3-1)\sqrt{\frac{a_0}{a_i}} - x^3\sqrt{\frac{a_0}{a}} + \frac{1}{2\left(\frac{\alpha}{2}-1\right)}\left[\left(\frac{a_0}{a_{int}}\right)^{\left(\frac{\alpha}{2}-1\right)} - 1\right]\right]$$

$$[2.58]$$

Before estimation, we must consider that the initial flaw, a_{int}, could be 3–10% less than an initial circular crack, a_0, and the slope, α, below threshold is much higher than 3 as it is used in the crack growth law [MAR 06, MAR 07].

$a_{int} = 0.9a_0$	$a_{int} = 0.94a_0$	$a_{int} = 0.97a_0$
$\alpha = 25$	$\alpha = 100$	$\alpha = 200$
$N_{int} \cong \dfrac{\pi E^2}{2(\Delta\sigma)^2} \times \dfrac{1}{10}$	$N_{int} \cong \dfrac{\pi E^2}{2(\Delta\sigma)^2} \times \dfrac{1}{5}$	$N_{int} \cong \dfrac{\pi E^2}{2(\Delta\sigma)^2} \times \dfrac{1}{10}$
$N_{int} \cong \dfrac{\pi E^2}{20(\Delta\sigma)^2}$	$N_{int} \cong \dfrac{\pi E^2}{10(\Delta\sigma)^2}$	$N_{int} \cong \dfrac{\pi E^2}{20(\Delta\sigma)^2}$

Table 2.1. *Approximate number of cycles below threshold*

The aim of the above formulations is to verify the influence of crack growing over the total number of cycles in the gigacycle regime, considering a range of transition sizes from small to large crack.

2.5. Example of fish-eye formation in a bearing steel

Here, we are going to illustrate a numerical example, which will be confronted with experimental results, obtained from a bearing martensitic steel (100C6) having internal failures at very high cycle fatigue [MAR 07]. An SN curve is shown in Figure 2.17. The fracture surface after 3×10^9 cycles is shown in Figure 2.18.

For low load ratio, $R \approx 0$, b/x^3, $x = 3$, $\alpha = 100$ and $a_{int} = 0.94a_0$

When $a > a_i$,

$$N_{fish\text{-}eye} = \frac{\pi E^2}{2(\Delta\sigma)^2}\left[1.2 + 26\sqrt{\frac{a_0}{a_i}} - 27\sqrt{\frac{a_0}{a_i}}\right]$$

When $a \leq a_i$, we must combine equations [2.55] and [2.52] to obtain:

$$N_{fish\text{-}eye} = \frac{\pi E^2}{2(\Delta\sigma)^2}\left[1.2 - \sqrt{\frac{a_0}{a_i}}\right]$$

For medium load ratio, $R \approx 0.4$, b/x^3, $x = 2$, $\alpha = 100$ and $a_{int} = 0.94a_0$

When $a > a_i$,

$$N_{fish\text{-}eye} = \frac{\pi E^2}{2(\Delta\sigma)^2}\left[1.2 + 7\sqrt{\frac{a_0}{a_i}} - 8\sqrt{\frac{a_0}{a}}\right]$$

When $a \leq a_i$, we use the same relationship obtained for $R \approx 0$

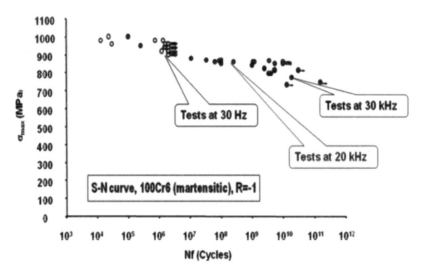

Figure 2.17. *A typical SN curve for bearing steel up to 10^{11} cycles. Note that there is no effect of frequency*

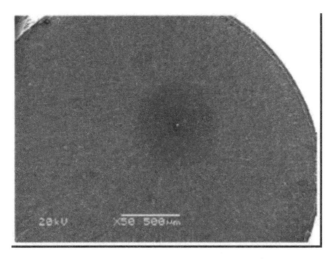

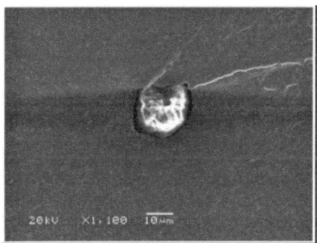

Figure 2.18. *Fish-eye formation in bearing steel from an
inclusion at 3×10^{9} cycles*

In Table 2.2, it is shown that the cyclic load ratio, R, does not have
a significant influence on the crack growth portion of life. In addition,
the variation of transition size from small to large crack does not have
a considerable effect on crack growth life, as compared to initiation
for GCF.

$\Delta\sigma$ (MPa)	N_{total}	$a_{0(\mu m)}$	$A_{(\mu m)}$	$R \approx 0$, $b/27$, $\alpha = 100$ and $a_{int} = 0.94a_0$			$R \approx 0$, $b/8$, $\alpha = 100$ and $a_{int} = 0.94a_0$		
				$a_i = 100\ \mu m$	$a_i = 125\ \mu m$	$a_i = 200\ \mu m$	$a_i = 100\ \mu m$	$a_i = 125\ \mu m$	$a_i = 200\ \mu m$
				$N_{fish\text{-}eye}$	$N_{fish\text{-}eye}$	$N_{fish\text{-}eye}$	$N_{fish\text{-}eye}$	$N_{fish\text{-}eye}$	$N_{fish\text{-}eye}$
860	2.97×0^9	12	413	5.25×10^5	4.36×0^5	2.78×0^5	2.12×10^5	1.88×10^5	1.45×10^5

Table 2.2. *Effect of R ratio and short crack transition on fish-eye growth in 100C6 steel*

In conclusion, GCF life is primarily due to initiation mechanisms, and again this shows that crack growth is not a significant portion of life.

2.6. Fish-eye formation at the microscopic level

First, at the macroscopic scale, the appearance (color) of the fish eye is not the same in an OM and SEM. Figure 2.10 shows the observations with an OM at low magnification and SEM at high magnification. These results are obtained on carburized low alloyed chromium steel, SAE 5120 [WEI 10].

The microstructure is martensite with retained austenite. In the optical micrograph, the crack initiation site appears as a dark point (called ODA), whereas in the SEM, its appearance is not so clear. Around the dark point, a white zone is present in the optical micrograph. In SEM observation, this zone is rather flat and circular (penny-shaped). Outside this white zone, we can see a wide fracture surface having a large radial ridge pattern in both optical and SEM micrographs. In this case, the crack initiation is located on an inclusion (mixed inclusion of Al, Si, Ca, Mg, Mn and S).

2.6.1. *Dark area observations*

The dark area zone corresponds to the crack initiation zone, which, in the GCF domain, can be an inclusion or a metallurgical structural heterogeneity (large grain)

As reported by Murakami, Sakai and Shiozawa, at high magnification (Figures 2.19 and 2.20), the dark area appears like a granular layer (called FGA or GBA) as seen in the SEM. The explanation of the formation of this FGA differs among researchers due to controversies that exist on the origin of this zone. Figure 2.20 shows crack initiation on a large grain (probably a large, soft, retained austenite grain) in SAE 5120. In this case, the FGA spreads over and around the entire grain as predicted by the Sakaï model.

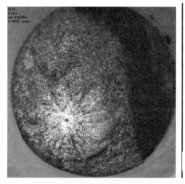

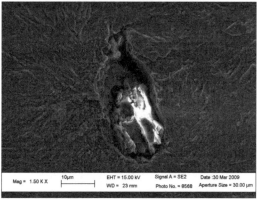

Figure 2.19. *Two different aspects of fish eye at low (OM) and high magnification (SEM). $N_f = 4.25 \times 10^7$ cycles*

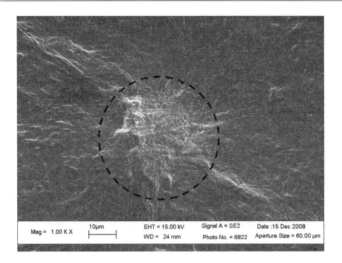

Figure 2.20. *Initiation from a large grain in SAE 5120.*
$N_f = 3.61 \times 10^7$ *cycles. The dotted line is for a_{int}*

2.6.2. *"Penny-shaped area" observations*

Whatever the crack initiation site (inclusion, "supergrain", porosity), the fracture surface eventually becomes circular (penny-shaped) around the initiation site. Figures 2.21 to 2.22 show examples of such behavior (as expected by the variation of the stress intensity factor around the periphery of the crack).

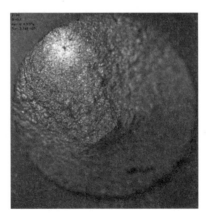

Figure 2.21. *Initiation from an inclusion in SAE 5120. $N_f = 2.34 \times 10^7$ cycles*

Figure 2.21 shows the area around the ODA/FGA for a specimen in a low alloyed chromium steel. This zone is flat as shown in Figure 2.22. The martensitic platelet microstructure is visible. It is expected that, in this zone, there is no crack closure.

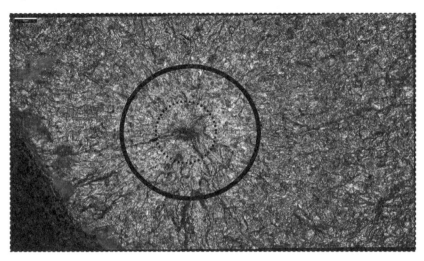

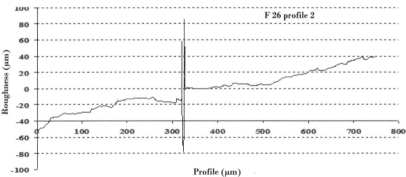

Figure 2.22. *Fish-eye formation from an inclusion. The radius of the penny-shaped area a_0 (82 μm) determined from its profile is related to the initiation stage and short crack growth up to the beginning of long crack growth, a_i [WEI 10]*

For this specimen, the fatigue life is 2.34×10^7 cycles since the theoretical number of cycles of propagation calculated with equation 2.54 is $N = 6 \times 10^5$ cycles. It means that the number of cycles for initiation is $N_i = 2.278 \times 10^7$ cycles.

According to Figures 2.22 and 2.23, the radius a_{int} (12 μm) is the size of the crack around the inclusion from which the propagation starts. The transition a_0 from short crack to long crack is determined by topography measurement on the fracture surface (Figure 2.18). In this case, $a_0 = 82$ μm. It is related to the so-called penny-shaped area. Finally, the size of the fish eye at the failure instability is $a = 273$ μm.

It must be pointed out that initiation of the crack appears to be the key phenomenon of GCF. The large number of cycles for initiation is shown both by fractography and by calculation. In Chapter 3, this will be confirmed by thermal measurement. From those observations, a schematic model of the fish-eye formation is shown in Figure 2.23; in agreement with it is the Paris–Hertzberg equation [2.52].

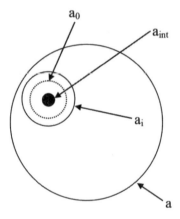

Figure 2.23. *Schematic of a fish-eye formation in gigacycle fatigue. ODA or FGA, a_0; transition long crack, a_{int}; crack instability, a_i; in agreement with the Paris–Hertzberg equation*

2.6.3. *Fracture surface with large radial ridges*

In such fish-eye zones, high magnification SEM observations show striations at 200 μm from the center (Figures 2.22, 2.24 and 2.25). It means that striations are occurring inside the fish eye, before the instability of the fish eye through the surface. All the observations and measurements were made on SAE 5120 steel for many specimens. Each picture at high magnification (5,000×) was cross-referenced with a picture at minor magnification (100×) to show the location of the striations with regard to the crack initiation site (Figure 2.24).

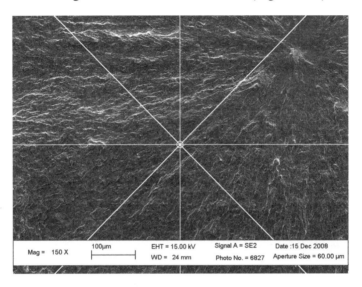

Figure 2.24. *Observation of striations at 200 μm from the center of the fish eye (Figure 2.22) in SAE 5120*

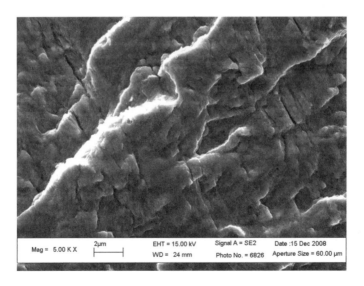

Figure 2.25. *Striations in a fish eye (detailed view of Figure 2.24)*

For each zone with striations, the mean distance between striations was measured and the associated ΔK was calculated using the formula:

$$\Delta K = 2\Delta\sigma/\pi\sqrt{\pi a}$$

Figure 2.22 shows the evolution of the striation spacing $\log(e)$ versus $\log(\Delta K)$.

The equation of the straight line is:

$$\log(e) = 2\log(\Delta K) - 10.97$$

The slope of the straight line is in good agreement with the CTOD model in which the striation spacing is proportional to the crack tip opening displacement, in other words a function of ΔK^2 as predicted by F. McClintock.

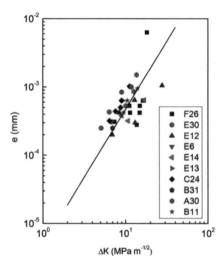

Figure 2.26. *Relation between the striation spacing and ΔK inside a fish eye. For a color version of the figure, see www.iste.co.uk/bathias/Fatigue.zip*

Until now, it was reported that inside the fish eye, vacuum is present. So, it is surprising that well-defined striations are present in the large radial ridge zone. A review of the environment effect by Petit and Sarrazin-Baudoux [PET 08] suggests that in vacuum, no striation appears. Nevertheless, striations in vacuum do appear possible but are less clearly defined.

Probably, the environment in the fish eye is not vacuum because during fabrication of the steel, an air partial pressure occurs or, as suggested by Murakami [MUR 02], if hydrogen is desorbed from non-metallic inclusions (at the initiation site, or inclusions on the crack path during the propagation), the environment is not a pure vacuum. However, the presence of striations in a fish eye is a good reference to determine the crack growth in stage 2 and to measure the crack growth rate inside the fish eye. It is of significance to note that, in SAE5120, the average crack growth inside the fish eye is of the order of 5×10^{-4} mm per cycle. It means that the number of cycles with striations inside the fish eye is of the order of few thousand. Therefore, the crack growth is very fast inside the fish eye.

2.6.4. *Identification of the models*

Detailed fractographic observations of the fish-eye fracture surface show several zones. The aim of this section is to relate these fractographic observations to our mechanical model (Figure 2.23).

The final zone of the fish eye from a_i to a in the mechanical model corresponds to the large radial ridges zone. For the crack length a, the fracture toughness of the material is reached, and fast and final fracture occurs. In the final zone (from a_i to a), striations are visible and the propagation is large crack growth in stage II.

In the mechanical model, the transition from short crack to long crack is steep at a crack length equal to a_i (Figure 2.23). For the fractographic observations, the "penny-shaped zone" in which the martensitic platelet microstructure is seen (for the low alloyed steel) is related to the short crack propagation where no crack closure is present. The crack length of the "penny-shaped zone" is called "a_{i-1}". Between the "penny-shaped zone" and the large radial ridge zone, a zone with small radial ridges is observed. At the end of this zone, called "a_{i-2}", sometimes striations are visible. So, as is foreseeable, the transition from short crack to long crack is not steep, but localized around a_i between a_{i-1} and a_{i-2} (as shown for one specimen in Figure 2.24).

In the mechanical model, the threshold is taken at $da/dN = b$ and $\Delta K_{eff}/E\sqrt{b} = 1$ and corresponds to a crack length initiation a_0. In the SEM observations, a fine granular zone is observed (probably the dark area of the optical observations) around the defect from which the crack initiation occurs. This zone (Figure 2.23) from a_{int} (defect size) to a_0 (FGA zone) is due to the formation of dislocation cells in grain(s) well oriented around the defect. This is also the beginning of the crack propagation in opening mode where the stress intensity factor is of the order of the threshold.

Quantitative measurements of the different zones were performed (on the low alloyed steel SAE 5120 tested at $R = 0.1$) every time it was possible [WEI 10, HUA 11]. The final crack length a of the fish eye was not considered, because the specimens studied have a carburized surface that induces a hard layer on the surface specimen, disturbing the true final crack length. For the other characteristic lengths, the following results are obtained:

– 40 µm $< a_{i-2} <$ 140 µm;

– 30 µm $< a_{i-1} <$ 70 µm ;

– 15 µm $< a_0 <$ 25 µm;

– 7 µm $< a_{int} <$ 19 µm;

where:

- a_{int} is the defect radius leading to the initiation;

- a_0 is the FGA radius due to polygonization and high density of dislocations;

- a_{i-1} is the beginning of the transition from small crack to large crack;

- a_{i-2} is the end of the transition from small crack to large crack;

- a is the fish-eye final radius.

For each crack measurement, ΔK ($\Delta K = 2\Delta\sigma/\pi\sqrt{\pi a}$) was calculated versus the number of cycles at failure. Whatever the number of cycles at failure, between 10^6 and 10^9 cycles, the stress intensity factor remains constant or decreases slightly around a circle determined by a_{int} and a_0, and at a_i, which is expected for the same material. Generally, it is found that the stress intensity factor is lower than the threshold between a_{int} and a_0. In accordance with the alloy and the size of the defect, a_{int} and a_0 are not constant, but the stress intensity factor at a_0 is of the order of the threshold corner of the Paris–Hertzberg model.

In the Paris–Hertzberg model, the threshold is predicted at $da/dN = b$ and $\Delta K_{eff}/E\sqrt{b} = 1$. This corner corresponds to the radius a_0, the fatigue crack threshold. It means that at the end of the FGA, the fatigue crack growth can be recorded using the Paris law but below a_0, fracture mechanics cannot be applied. This stage is devoted to initiation for 99% of the total fatigue life.

Taking E = 210,000 MPa (for steels) and b = 0.3 nm, the value of ΔK_{eff} at the thresholdis 3.64 MPa.

Assuming that:

$$\Delta K = \Delta K_{eff}/(1 - R)$$

at $R = 0.1$, which gives $\Delta K = 4$ MPa (This value corresponds to ΔK_{a0} and is in very excellent agreement with the threshold value (4 MPa$\sqrt{m}$)) given by equation [2.54].

The transition from short crack to long crack is obtained for a mean stress intensity factor ΔK_{ai1} equal to 6.6 MPa$\sqrt{m}$. The difference between ΔK_{ai1} and ΔK_{a0} (the factor x in the mechanical model) is approximately 2.6 MPa$\sqrt{m}$. This factor x has been observed at a maximum around 3 for low load ratios ($R = 0$), and is in very good agreement with the mean experimental value obtained on the low alloyed steels [MAR 06].

2.6.5. *Conclusion*

The different zones of a fish eye in the GCF domain have been characterized using optical and scanning microscopy. The fractographic results show zones in agreement with our mechanical model:

– A dark area zone (a_{int} to a_0) due to dislocation cell formation or polygonization.

– A penny-shaped zone (a_0 to a_{i-1}, crack growth in stage I). Whatever the crack initiation site (spherical or elongated inclusion, supergrain, pore), the fracture surface becomes circular (penny-shaped) around the initiation site.

– A zone with small radial ridges (a_{i-1} to a_{i-2}) corresponding to the short to long crack transition centered on a_i.

– A zone with large radial ridges (a_i to a, crack growth in stage II). In this zone, the fatigue crack propagation produces striations for which the mean distance between striations is a function of ΔK^2, in good agreement with the CTOD model.

– It is also shown that the theoretical fatigue crack threshold is always related to a_0, larger than the defect radius. It means that the crack cannot grow from the first cycle, according to a Paris law. A nucleation is necessary before crack growth.

2.7. Instability of microstructure in very high cycle fatigue (VHCF)

Through our own results and from the literature, it is observed that the microplasticity in the GCF is induced more than dislocation sliding. Sometimes, phase transformation, refining of the grain, twinning and instability of the yield point occur even when the loading is small during a very high number of cycles. Some typical observations in GCF at high frequency are summarized below:

– *Effect of austenite in stainless steels*: in austenic stainless steels, the austenite is not stable in the GCF regime even if the plastic deformation is theoretically very small. There is a large thermal dissipation.

– Effect of residual austenite in martensite: when the rate of austenite approaches 10% in martensitic steels, a large thermal dissipation is observed at the beginning of the test followed by a high temperature rise when the fish eye propagates. The thermal dissipation associated with the microstructural instability is a limitation in ultrasonic fatigue.

– Instability of titanium alloys: more surprising is the instability of some Ti alloys at cryogenic temperature at ultrasonic frequency.

– Possible effect of the dislocation mobility: in low-carbon steels, the mobility of dislocations is affected by interstitial atoms, depending on the strain rate and the grain size. Several types of instability are observed in monotonic loading such as Luder's bands, Portevin–Le Chatelier bands or Neuman's bands, twinning, etc. It seems useful to consider similar effects in ultrasonic fatigue at high frequency.

One of the most interesting behaviors for practical applications is the instability of martensite or bainite in high-strength steels. Depending on the chemical composition, heat treatments and processing, several types of microstructure transformation are observed around the defect in the center of the fish eye. Several examples are given below. In ferrite-perlite steel (D38), in the center of a fish eye, the initiation starts from a colony of perlite, the so-called supergrain, (Figure 2.27) but in a martensitic steel, the initiation starts generally from an oxide inclusion, as shown in Figure 2.28. This is typical of local plasticity. However, in high-strength steel, in other words when the UTS is more than 900 MPa, the initiation of the crack in the center of the fish eye is sometimes more complicated as shown in Figure 2.29.

In high carbon content and in high strength steel, a transformation of the microstructure (martensite or bainite) starting from the inclusion seems to occur, in relation with the stress concentration and the stress field (Figure 2.29).

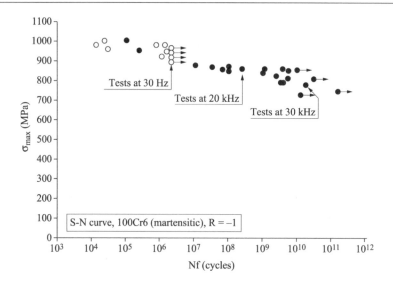

Figure 2.27. *Initiation from perlite colony in D38 carbon steel*

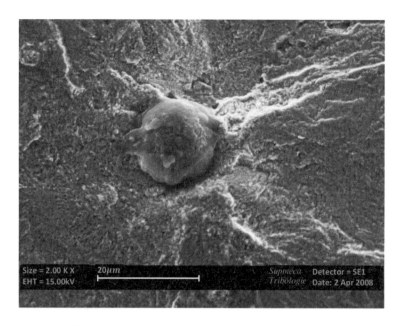

Figure 2.28. *Initiation from an oxide in martensitic steel*

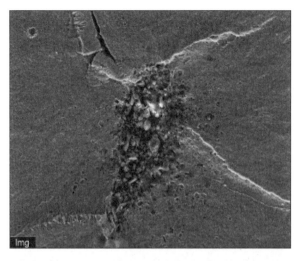

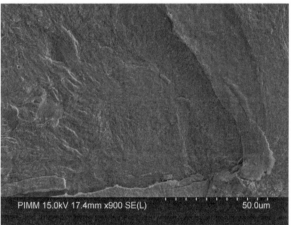

Figure 2.29. *Initiation with phase transformation and grain boundary cracking starting from an oxide*

It is difficult to understand this phase transformation that appears at the microscopic scale in a radius of 200 μm around an inclusion. From the SEM observation, the following can be said:

– Some wings (two to four) may occur around the inclusion similar to the butterfly wings observed in rolling contact fatigue. Microcracks are observed along the boundary between the wings and the matrix.

Some wings are broken in brittle appearance (Figure 2.29). In accordance with the rolling contact damage, it seems that a phase transformation or a grain refining occurs (Figures 2.29–2.31). In both cases, it is reasonable to assume that the microstructure of the wings should be a nano-ferrite phase, as shown by Grabulov et al. [GRA 07] in rolling fatigue. Figure 2.30 compares the initiation of a fatigue crack in rolling fatigue and in push–pull fatigue for a same bearing steel. Indeed, the feature is very similar.

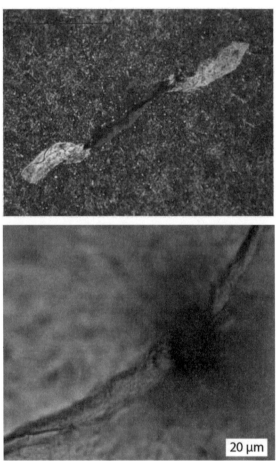

Figure 2.30. *Comparison between wings in rolling contact fatigue and push–pull gigacycle fatigue in a bearing steel*

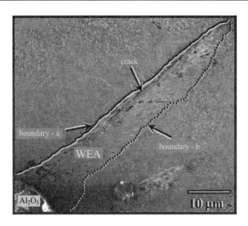

Figure 2.31. *Rolling fatigue in bearing steel. SEM image of a butterfly crack formed around Al_2O_3 inclusion indicates the white etching area (WEA) "butterfly wing" having two boundaries: one containing a crack, denoted as "boundary-a", and the other without a crack denoted as "boundary-b" [GRA 07]*

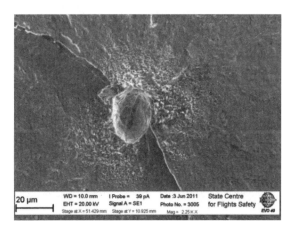

Figure 2.32. *Gigacycle fatigue in martenstic steel. SEM image of a butterfly crack starting from an oxide inclusion with two butterfly wings along austenitic grain boundary in martensitic steel failed in push–pull loading. Around the inclusion, an FGA is present (A. Nikitin)*

– Figure 2.31 shows a butterfly wing after rolling, fatigue having two boundaries: one containing a crack and the other without a crack. Figure 2.32 shows the same damage but in GCF. In Figure 2.32, the butterfly wing is related to the direction of grain boundary [SHA 12].

– It is of significance to note the relation between the wings, the oxide at the center of the fish eye and the former austenitic grain boundaries, which is clear in Figure 2.33. To prove this relation, a fish eye at the center of a fish eye in a high-strength steel specimen was observed with an SEM/EBSD. The results are shown in Figure 2.33.

– It is shown that the oxide inclusion, from which the initiation starts, is located at a triple point of former austenitic grains. The wings or the phase transformation appears along these grain boundaries producing internal stress and cracking along the wings for a length of approximately 200 μm. It was not possible to confirm the crystallography of the wings. But according to the results presented at VHCF4 [GRA 07], it should be a transformation of martensite in ferrite.

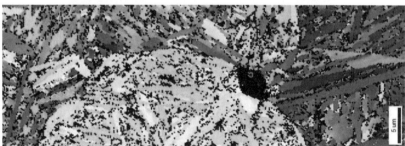

Figure 2.33. *Relation between an oxide inclusion located at a triple point and three wings along the prior austenitic grain boundary (courtesy of D. Field). For a color version of the figure, see www.iste.co.uk/bathias/Fatigue.zip*

This approach proves that the initiation of a fish eye in GCF is not only due to PSB, polygonization, with or without hydrogen effect. The grain boundary, the interaction between defect and grain boundary, and sometimes the phase transformation or the refining of the microstructure are involved in a complex process up to the formation of a crack in opening mode. The formation of butterfly wings in martensitic or bainitic steels shows the effect of the shear stress field at the beginning of the fish-eye initiation from an inclusion, see [SHA 12].

2.8. Industrial practical case: damage tolerance at 10^9 cycles

After the discussion of this phenomenological approach, it is of significance to discuss some technological requirements facing the GCF design. We have chosen a powder metallurgy nickel base alloy (N18) with and without seeding of inclusions used for jet turbine disks [BAT 10]. Such a component is designed not only in LCF but also in vibration, that is to say in GCF. The chemical composition of N18 nickel base alloy is:

$$Cr_{11,5} Co_{15,5} Mo_{6,5} Al_{4,3} Ti_{4,3} Nb_{1,5} HF_{0,5}$$

The principal mechanical properties are: σ_y = 1,050 MPa, UTS = 1,500 MPa.

To reveal the effect of defects, a pollution with ceramic inclusions (80–150 µm diameter) was made using 30,000 inclusions for 2.20 pounds of alloy. A comparison is made between N18 alloy with and without inclusions. Nevertheless, it is observed that when the rate of ceramic particles increases, the porosity is also increased, in other words a competition exists between particles and pores for crack nucleation (Figure 2.34).

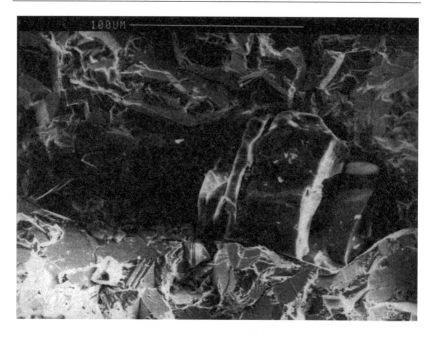

Figure 2.34. *Defect related to gigacycle initiation in N18 alloy [BON 97]*

2.8.1. *Fatigue threshold in N18*

Figures 2.35 and 2.36 show the results at high frequency to $R = -1$. We can see that the threshold is smaller at high temperature than at ambient temperature. Normally, we could wait for a fall of threshold with the increase in the temperature. But in Figure 2.35, the threshold is smaller at 400°C than at 650°C and 750°C. The curves at 400°C, 650°C and 750°C cut the vicinities 10^{-5} mm/cycle. The observed gaps are explained by the phenomenon of oxidization at the bottom of the crack. On the crack surface of the sample used in our tests, the oxidization at 650°C and 750°C was observed. At high temperature, normally the propagation crack rate increases with the temperature. But the oxidization could slow down propagation to the neighborhood threshold to a small load when the temperature is rather elevated.

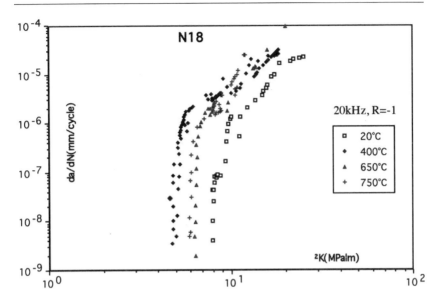

Figure 2.35. *Fatigue threshold for N18 at R = −1. For a color version of the figure, see www.iste.co.uk/bathias/Fatigue.zip*

It is found that the ΔK thresholds for $R = -1$, $R = 0$, $R = 0.8$ are, respectively, 5.5 MPa√m, 8 and 4.2 at 450°C, which is the basic temperature of a turbine disk.

2.8.2. *Fatigue crack initiation of N18 alloy*

Figure 2.36 shows the fatigue results on N18 nickel base alloy at 20 kHz and for a different ratio R. The specimens are polished before testing. There is no horizontal asymptote on the SN curve between 10^6 and 10^7 cycles. Between 10^6 and 10^9 cycles, a flat SN curve with a uniform slope is observed. It means that the fatigue limit defined as an asymptotic value of the stress up to 10^6 is an incorrect concept in this case. It is found that for long-life range, the initiation of the crack starts inside the specimen from a defect. It means that the number of cycles for initiation depends on the size of the defect, the location of the defects and also on low crack growth rates in vacuum before the internal crack collapses at the surface of the specimen.

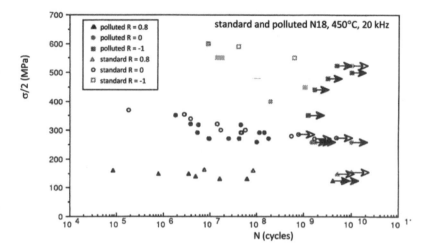

Figure 2.36. *SN curves for N18 at 450°C*

At $R = -1$, the fatigue strength at 10^9 cycles of the standard material is 525 MPa. But in presence of inclusions, the fatigue limit can drop to 400 MPa.

The curves of standard and polluted N18 with $R = 0$ are shown in Figure 2.36. Initiation sites are meanly porosities for the standard N18 and ceramics inclusions for polluted material inside the sample and not on the surface. The scatter of the data for seeded material is higher than the standard because the size, shape and interfaces with a matrix of porosities are more constant than inclusions. But it is outstanding that fatigue limits at 10^9 cycles are not so different (270 MPa for standard material and 260 MPa for seeded material), although the average size of porosities and inclusions are quite different (23 μm for standard material and 98 μm for seeded material).

It is also observed that for a standard SN curve, the fatigue limit at 10^9 cycles is 75 MPa lower than the fatigue limit at 10^6 cycles.

At $R = 0.8$, the fatigue strength of seeded material is 25 MPa lower than standard material (125 MPa for seeded material and 150 MPa for standard material). A large mean stress value increases the ceramic inclusion effect.

2.8.3. *Mechanisms of the GCF of N18 alloy*

According to our own observations, the GCF crack initiation essentially occurs inside the sample and not at the surface as is observed for some shorter lives. So, we can predict the crack initiation from the existing defect, which nucleates a fish eye.

As an example, we consider the specimens failed at 291 MPa and $R = 0$ (Figure 2.34) after 10^8 cycles. The average size of the defect in the middle of the fish eye is 100 µm, for a fish-eye diameter of 1,200 µm. The maximum size of the defect (more or less an ellipse) is 165 µm.

According to equation [2.54], it is found that N_f, the number of cycles of propagation inside the fish eye, is approximately 7.3×10^5 cycles. This means that the number of cycles of initiation is much more important than the propagation of the short crack. It is not reasonable to consider that a crack can propagate from an inclusion without nucleation in the gigacycle regime.

Now, considering the threshold concept, the initial ΔK_i at the tip of the defect is found between 3.28 and 4.21 MPa/m, depending on the average or maximum size of the defect. This is less than or equal to the fatigue threshold of a short crack in such alloys at 450°C (Figure 2.35). This means that the crack growth rate is of the order of 10^{-6} mm/cycle or higher, following the results of Figure 2.35. In these conditions, the number of cycles of propagation in a fish eye of 600 µm radius should be approximately 6×10^6 cycles in good agreement with N_f calculated above.

In the gigacycle regime of N18 nickel-based powder metallurgy (PM) alloy used for a disk, the fatigue strength is related to the dislocations sliding around defects more than short crack growth. Comparing the effect of inclusions and porosities in the N18, it is interesting to point out the following:

The role of the inclusions is sometimes in competition with the role of porosities when the load ratio R is equal to –1 or 0. The scattering of the results for $R = 0$ in an N18-polluted inclusion is important, since for an N18 standard, the scattering is low.

It is very remarkable that when $R = 0$, the resistance to the GCF at 450°C is approximately 250 MPa in the N18 with or without inclusions, the porosity being dominant.

On the contrary, when the static stress of the fatigue cycle is very high, for $R = 0.8$, the role of inclusions becomes preponderant. Without inclusions, the N18 fatigue strength, at 450°C, is 155 MPa at 10^9 cycles. However, it is 125 MPa with inclusions.

From an engineering point of view, it is clear that the damage tolerance concept is not the best approach for designing a turbine disk made by powder metallurgy, loaded in vibration. In the gigacycle regime, when a crack is detected by a conventional non-destructive testing (NDT) method, it is too late, the disk has failed. A prediction of the fish-eye initiation is mandatory under vibration loading.

2.9. Bibliography

[BAT 10a] BATHIAS C., "Influence of the metallurgical instability on the gigacycle fatigue regimen", *International Journal of Fatigue*, vol. 32, no. 3, pp. 535–540, March 2010.

[BAT 10b] BATHIAS C., PARIS P.C., "Gigacycle fatigue of metallic aircraft components", *International Journal of Fatigue*, vol. 32, pp. 894–897, 2010.

[BAT 13a] BATHIAS C., "Coupling effect of plasticity, thermal dissipation and metallurgical stability in ultrasonic fatigue", *International Journal of Fatigue*, August 2013.

[BAT 13b] BATHIAS C., FIELD D., ANTOLOVICH S., *et al.*, "Microplasticity, micro damage, microcracking in ultrasonic fatigue", *ICF13 proceedings/symposium VHCF Proceedings*, June 2013.

[BON 97] BONIS J., Gigacycle fatigue of nickel base alloys at high temperature, PhD Thesis, Univeristy of Orsay, Paris, 1997.

[CHO 13] CHONG W., PhD report, University Paris Ouest, June 2013.

[GRA 07] GRABULOV A., ZIESE U., ZANDERGEN H.W., "Investigation of microstructural changes within white etching area under rolling contact fatigue", *VHCF 4 Proceedings, TMS 219–226*, 2007.

[HER 93] HERTZBERG R.W., *Int. Journal of Fracture*, vol. 64, no. 3, p. 135, 1993.

[HER 95] HERTZBERG R.W., *Mat.Sci. Eng. A190*, vol. 25, 1995.

[HER 12] HERTZBERG R.W., "Deformation and fracture of engineering materials", John Wiley, 2012.

[HUA 11] HUANG Z., WAGNER D., BATHIAS C., *et al.*, "Subsurface crack initiation and propagation mechanisms in the gigacycle fatigue", *Acta Materialia*, pp. 136–145, 2011.

[KLE 65] KLESNIL M., LUKÁŠ P., "Dislocation arrangement in the surface layer of α-iron grains during cyclic loading", *Journal of the Iron and Steel Institute*, pp. 1043–1047, October 1965.

[MAR 06] MARINES-GARCIA I., PARIS P.C., TADA H., *et al.*, "Fatigue crack growth from small to large cracks in gigacycle fatigue with fish-eyes failure", *9th International Fatigue Congress*, Atlanta, GA, May 2006.

[MAR 07] MARINES-GARCIA I., PARIS P.C., TADA H., *et al.*, "Fatigue crack growth from small to large cracks in gigacylce fatigue with fishes eyes failure", *International Journal of Fatigue,* vol. 29, p. 2072, 2007.

[MUG 76] MUGHRABI H., HERZ K., STARK X., "The effect of strain-rate on the cyclic deformation properties of α-iron single crystals", *Acta Metallurgica*, vol. 24, pp. 659–668, 1976.

[MUG 06] MUGHRABI H., "Specific features and mechanisms of fatigue in the ultrahigh-cycle regime", *International Journal of Fatigue,* vol. 28, pp. 1501–1508, 2006.

[MUR 02] MURAKAMI Y., *Metal Fatigue: Effects of Small Defects and Nonmetallic Inclusions*, Elsevier, 2002.

[NAK 04] NAKASONE Y., HARA H., "FEM simulation of growth of fish-eye cracks in very high cycle fatigue of high strength steel", *VHCF Proceedings,* pp. 40–48, September 2004.

[PAR 04] PARIS P.C., MARINES-GARCIA I., HERTZBERG R.W., *et al.*, "The relationship of effective stress intensity factor, elastic modulus and Burgers-vector on fatigue crack growth as associated with "Fish Eye" gigacycle fatigue phenomena", *Proceedings of the 3rd International Conference on Very High Cycle Fatigue (VHCF-3)*, Kyoto, Japan, 16–19 September, 2004.

[PAR 99] PARIS P.C., TADA H., DONALD J.K., *Int Journal Fat. 21*, vol. S35, p. 125, 1999.

[PET 08] PETIT J., SARRAZIN-BAUDOUX C., "Effet de l'environnement", in BATHIAS C., PINEAU A. (eds), *Fatigue des matériaux et des structures 2*, Lavoisier, Paris, pp. 149–192, 2008.

[PIN 10] PINEAU A. (ed.), "Low cycle fatigue", *Fatigue of Materials and Structures*, John Wiley and Sons, New York, pp. 113–177, 2010.

[SHA 12] SHANYAVSKIY A., NIKITIN A., CHONG W., *et al.*, "An understanding of crack growth in VHCF from an internal inclusion in steel", *Crack paths 12 Proceedings*, Gaeta, 2012.

[SHI 04] SHIOZAWA K., NISHINO S., SHIBATA N., "Efffect of tempering temperature on super long fatigue behaviour of low alloy steel", *VUCF 3 Proceedings*, pp. 609–616, 2004.

[SUG 04] SUGETA A., UEMATSU Y., JONO M., "An effect of load variation on high-cycle fatigue behavior of high strength steel", *VHCF 3 Proceedings*, pp. 177–185, September 2004.

[TAD 00] TADA H., PARIS P.C., *Handbook of Stress Concentration*, 2000.

[WEI 10] WEI WEI D., Couplage Thermomécanique sous sollicitations monotone et cyclique, PhD Thesis, University of Paris, 2010.

Heating Dissipation in the Gigacycle Regime

In Chapter 3 the coupling between the temperature increase, the heating dissipation, the micro plasticity and the damage is discussed. At 20 kHz, the temperature of the specimen becomes a crutial parameter.

3.1. Temperature increase at 20 kHz

Depending on the applied stress level, a heating dissipation is always observed at 20 kHz if no artificial cooling is applied. This why it is recommended to use a piezoelectric machine, only in the gigacycle fatigue, beyond a fatigue life of 10^6 cycles. Indeed, it depends on the computer control of the machine, assuming that the frequency and the amplitude of the stress are constant during the test.

At the beginning of the test, the temperature rapidly increases, followed by stabilization (Figure 3.1). The higher the applied stress, the larger the increase in temperature and the more significant the energy dissipated. At the end of the test, the temperature increases very rapidly until the fracture. This sudden increase allows determining the cycle number at crack initiation and proves that more than 90% of the total life is devoted to the initiation of the crack. The portion of the total life devoted to the initiation increases with the number of cycles to failure. When the initiation is in the subsurface and leads to a fish-eye formation, the simplified Paris model allows us to predict the number of cycles for fish-eye formation in good agreement with the temperature measurements. Thus, a coupling effect exists between thermal dissipation, plastic deformation and damage in the gigacycle regime. The idea of relating fatigue to intrinsic dissipation seems to be highly relevant. Since 1921, Moore and Kommers [MOO 21] have suggested predicting fatigue using the temperature increase. However, their procedure based on the thermocouples did not permit us to thoroughly establish the temperature increase as being identical to the fatigue damage under repeated stress. In the 1980s, infrared cameras permitted us to evaluate the temperature increase under cyclic stress and to describe the damage process up to the location of the initiation (see the research of Luong [LUO 98], Chrysochoos and Louche [CHR 00] and Hervé [HER 13]).

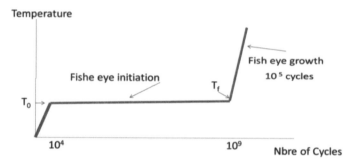

Figure 3.1. *Schematic evolution of the temperature during gigacycle testing*

For a gigacycle fatigue regime, a special approach is suggested. The temperature field on the surface of the specimen is measured using a non-intrusive measurement technique by an infrared pyrometer. In the test, an advanced, high-speed and high-sensitivity infrared CEDIP camera from Infrared Systems (made up of a matrix of 320×240 Mercury Cadmium Telluride detectors) was used to record the temperature changes during ultrasonic fatigue tests. The spectral range of the camera lies between 3.7 and 4.9 µm. This method allows the visualization of temperature cartography with very good time and spatial resolutions. In this study, the spatial resolution is 0.17 mm, and the aperture time can vary between 100 µs (lower than 1 cycle period) and 1,500 µs (approximately 30 cycles). The refresh time of the camera is between 0.83 and 100 Hz. The pyrometer was calibrated on a blackbody reference. In pyrometry, the error in temperature is often related to uncertainties on the emissivity of the surface. To eliminate this problem, the specimen surface was covered with a strongly emissive black paint. A schematic evolution of the temperature during a fatigue test with a fatigue life between 10^6 and 10^{10} cycles is shown in Figure 3.1. It is observed that between T_0 and T_F, the increase in temperature is very small. This plateau is related to the formation of the fine granular area (FGA) or to the initiation process. The effect of microstructure and of phase transformation is significant.

Some results are summarized below for different alloys when ultrasonic testing is carried out without any artificial cooling, in order to emphasize the relation between heating dissipation and microstructural effects. The temperature T_0 of the plateau, before propagation of the crack, is recorded for the same gigacycle regime (10^8–10^9 cycles) [CLA 13].

Spherical graphite cast iron	60°C at $R = -1$
Spring steel	Less than 100°C at $R = -1$
4240 steel	Less than 80°C at $R = -1$
Low-carbon steel	Less than 80°C at $R = -1$
AISI 5120 without γ residual	Less than 60°C at $R = -1$
AISI 5120 with γ residual	235°C at $R = -1$
Aluminum alloys	35°C at $R = -1$
TA4V6	Less than 80°C at $R = -1$
316 stainless steel	300°C at $R = -1$
22Cr4Ni stainless steel	300°C at $R = -1$
Ni-based Alloys	Less than 80°C at $R = -1$
Armco iron	60°C at $R = -1$

From the experimental observations reported above, it appears that the temperature of the specimen loaded at 20 kHz is strongly dependent on the microstructure and the stability of the different phases. In stainless steels, the martensitic transformation of austenite or the mechanical twinning induced a temperature increase.

As an example, Table 3.1 presents the chemical composition and Figure 3.2 shows the microstructure of 22Cr 4Ni steel before loading at 20 kHz and after loading in the gigacycle regime. The microstructure is transformed by twinning.

C	Si	Mn	Cr	Ni	W	Nb	N
0.53	Max 0.25	9.50	21.00	3.80	1.00	2.20	0.50

Table 3.1. *Chemical composition of 22Cr-4Ni steel*

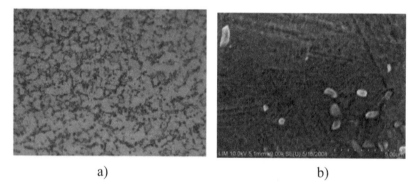

a)   b)

Figure 3.2. *Microstructure of a 22Cr-4Ni stainless steel before fatigue a) and after b) ultrasonic fatigue*

However, in many other alloys, such as martensitic steels, carbon steels, aluminum, titanium, nickel alloys, the increase in temperature due to the frequency is negligible from an engineering point of view.

3.2. Detection of fish-eye formation

To improve quality of materials, it is necessary to understand why they can fail at 10^9 cycles under small elastic loading and why the initiation may occur. Three stages should be distinguished: microplasticity, initiation and propagation. From each stage, a typical temperature increase should be observed. Obviously, during damping, the heating source is not the same as that during crack propagation. The initiation phase in the gigacycle fatigue can be described in terms of a local microstructurally irreversible portion of the cumulative plastic strain. When crack initiation appears, a short crack propagates followed by a long crack. In all cases, a cyclic plastic zone around the crack occurs, leading to a high elevation of thermal dissipation. So, the recording of the surface temperature of the sample during the test must allow us to follow the fatigue damage and determine the number of cycles at the crack initiation.

Using an advanced infrared imaging camera during gigacycle tests inside a piezoelectric fatigue system, the temperature evolution measured on the surface specimen during the tests is analyzed before and during crack initiation (with special attention to the number of cycles at initiation; Figure 3.3).

Figure 3.3. *Piezoelectric fatigue machine with an infrared thermal camera*

Figure 3.4 shows an increase in temperature, in a high-strength steel, just at the end of the test, which corresponds to the thermal dissipation during crack growth in the specimen.

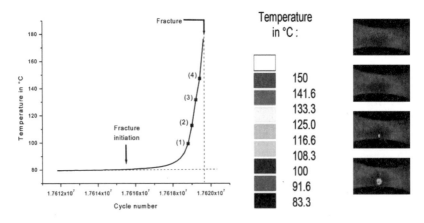

Figure 3.4. *Increase of temperature at the end of a gigacycle test of 4240 steel. For a color version of the figure, see www.iste.co.uk/bathias/Fatigue.zip*

In the same material, the temperature increases depending on the maximum stress amplitude for a given number of cycles. It should be noted that this quick increase in temperature at the beginning of the test is followed by a stabilization corresponding to a balance between the mechanical energy dissipated into heat and the energy lost by convection and radiation at the specimen surface and by conduction inside the specimen. As the crack growth occurs, we see that the temperature increases rapidly.

From the temperature recording, the number of cycles at crack initiation can be determined at this end of the plateau (Figure 3.4). As the crack begins to grow in opening mode, we see that the temperature increases rapidly due to a cyclic plastic zone, ahead of the crack, and leads to high thermal dissipation. There are two difficult problems to experimentally detect the initiation in the gigacycle regime: the great number of cycles and the location beneath the surface. However, the surface temperature recording is a very interesting technique to accurately determine the number of cycles at the initiation of the crack. In such tests, the number of cycles at crack initiation is more than 90% of the total number of cycles beyond 10^7 cycles, in agreement with the Paris–Herzberg derivative relation.

3.3. Experimental verification of N_f by thermal dissipation

Another experimental verification of the calculation of N_f, the number of cycles devoted to fish-eye growth, by thermal dissipation is underway, in a cast aluminum alloy, where the initiation is easier to determine than in steel.

The thermal dissipation in aluminum is shown in Figures 3.5 and 3.6. The location of gigacycle fatigue initiation is closer to the surface in cast aluminum than in steel, which makes it easier to be observed. Figure 3.5 shows that the temperature T_0 of the plateau depends on the stress level. At $R = -1$, for $\sigma_{max} = 35$ MPa, the increase in temperature is only 3°C but for $\sigma_{max} = 55$ MPa, the increase in temperature is more than 10°C with a total fatigue life of 8.1×10^7 cycles. When the crack begins to grow in opening mode, the temperature is 15°C.

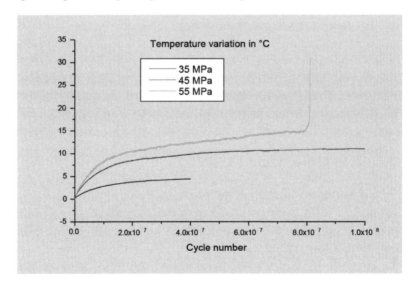

Figure 3.5. *Temperature increase for different fatigue life in a cast Al alloy. For a color version of the figure, see www.iste.co.uk/bathias/Fatigue.zip*

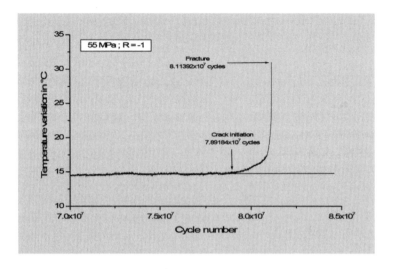

Figure 3.6. *Detection of the transition between initiation and propagation in cast Al alloy*

For a rough approximation, the previous derived relation [2.54] is used to estimate growth inside a fish eye:

$$N_f = \frac{\pi E^2}{2(\Delta\sigma)^2}$$

where:

E = 72560 MPa and $\Delta\sigma$ = 55 MPa

The theoretical number of propagation inside the fish eye N_{fi} is equal to 2.73×10^6 cycles. According to the temperature measurement shown in Figure 3.6, the experimental N_f is 2.2×10^6 cycles. Many other results show similar agreement.

Thus, it is possible to follow the temperature of the specimen during an ultrasonic fatigue test and to detect the location of the heating source and the location of the crack initiation, if any. The conclusions for steels and for aluminum alloys are the same.

3.4. Relation between thermal energy and cyclic plastic energy

To understand the propagation of the crack in the fish eye, the temperature field, the thermal dissipation after T_f and the plastic deformation at the crack tip have been studied in order to determine any relationship between these phenomena. A model of the thermal effects is associated with the crack propagation [RAN 10].

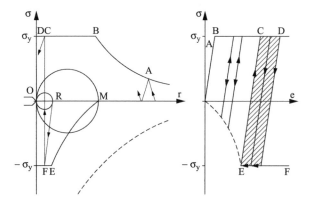

Figure 3.7. *Reverse plastic deformation (OR) at a fatigue crack tip, according the McClintock model*

According to Irwin and McClintock [BAT 10], it is supposed that the cyclic plastic zone (OR) in plane strain conditions at the fatigue crack tip (Figure 3.7) is given by the following relation:

$$r_R = \frac{\Delta K^2}{24\pi\sigma_y^2} \tag{3.1}$$

where K is the stress intensity factor and σ_y is the cyclic yield stress.

It is shown by Ranc *et al.* [RAN 10] that the plastic deformation energy is given by:

$$\mathcal{E} = \eta \frac{\Delta K^4}{24^2 \pi^2 \sigma_y^4} = \eta \frac{a^2 \Delta\sigma^4}{36\pi^4 \sigma_y^4} = \eta \frac{a_0^2 \Delta\sigma^4}{36\pi^4 \sigma_y^4 \left(1 - \dfrac{bft}{2a_0}\right)^4} \tag{3.2}$$

η : material coefficient;

$\wp_0$: plastic deformation energy per unit of time and of length;

f: frequency of the test (f = 20,000 Hz);

The dissipated power per unit crack length is given by:

$$P = \varepsilon f = \eta \frac{f a_o^2 \Delta \sigma^4}{36 \pi^4 \sigma_y^4 (1 - \frac{bft}{2a_0})^4} = \frac{P_o}{(1 - \frac{t}{t_c})^4} \qquad [3.3]$$

where $t_c = \dfrac{2a_0}{bf}$ and $P_0 = \eta \dfrac{f a_0^2 \Delta \sigma^4}{36 \pi^4 \sigma_y^4}$

ρ: density (kg/m^3);

C: specific heat (J/kg °C);

T: temperature (°C);

$\wp(t)$: dissipated plastic power (W);

λ: thermal conductivity (W/m °C);

r_c: radius of the plastic zone (O$_c$, x$_c$, y$_c$);

δ: the Dirac function;

Δ : Laplacien operator;

$a(t)$: crack radius at t (µm).

Starting from the heat transfer equation:

$$\rho C \frac{\partial T}{\partial t} = \frac{\wp(t)}{2} \delta(r_c - a(t)) \delta(z) + \lambda \Delta T$$

it is found that the evolution of the non-dimensional temperature during the fatigue crack growth is given by:

$$\frac{\partial T^*}{\partial t^*} = \frac{\delta^*\left(r_c^* - a^*\right)\delta^*\left(z^*\right)}{\left(1 - \dfrac{t^*}{t_c^*}\right)^4} + \Delta^* T^* \qquad [3.4]$$

As an example, a high-strength steel SAE 5120 (Martensite) loaded at 20 kHz, 410 MPa, $R = 0.1$ failed at 2.338×10^7 cycles. The fish-eye initiation from an inclusion is shown in Figure 3.8. The transition short crack/long crack is shown by a circular crack with a diameter of 160 μm. Figure 3.9 shows the increase in temperature during crack propagation inside the fish eye versus the number of cycles, and Figure 3.10 shows the curve calculated from relation [3.4]. The comparison between the results of the numerical simulation and the experimental determination of the temperature increase, during gigacycle fatigue, is in good agreement. It is shown that the nucleation of the crack is more than 99% of the fatigue life in the range of 10^7–10^8 cycles. The thermal calculation of Ranc *et al.* [RAN 10] shows that a temperature variation of 0.07°C corresponds to a small fish-eye radius of 20 μm. But when the crack grows, the temperature increases quickly depending on ΔK^4, in relation with the microstructure of the cyclic plastic zone, between a_i and a_f.

Figure 3.8. *Fish-eye initiation in SAE 5120. Failure at 2.338×10^7 cycles; $R = 0.1$; specimen F26*

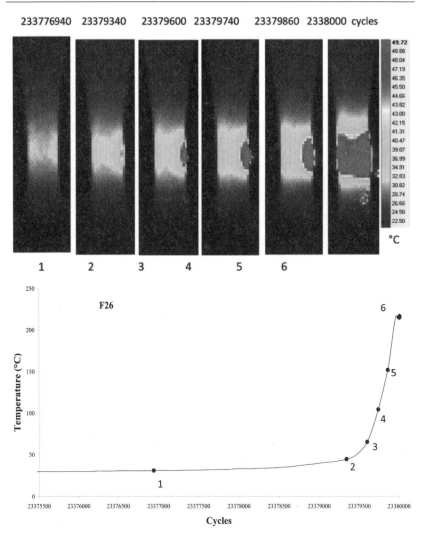

Figure 3.9. *Specimen F26. Increase in temperature after T_f (45°C) and thermo image of the fish-eye propagation ($N_f = 2.3 \times 10^4$ cycles) (Wei Wei DU). For a color version of the figure, see www.iste.co.uk/bathias/Fatigue.zip*

Finally, another example is shown in Figure 3.5 for a bearing steel 51200 (martensite + retained austenite) where a phase transformation occurs during fatigue loading with a high temperature increase during the plateau T_0.

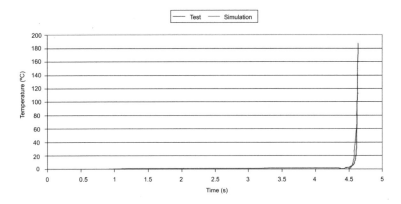

Figure 3.10. *Comparison between model and experiment for specimen F26. For a color version of the figure, see www.iste.co.uk/bathias/Fatigue.zip*

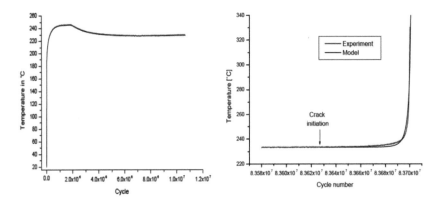

Figure 3.11. *Experiment and model for increase of temperature T_0 and T_f in 51200 bearing steel during gigacycle fatigue For a color version of the figure, see www.iste.co.uk/bathias/Fatigue.zip*

3.5. Effect of metallurgical instability at the yield point in ultrasonic fatigue

As mentioned earlier, the temperature increase depends on the metallurgical structure stability.

It is of significance to discuss the temperature of the plateau T_0 before crack propagation in the gigacycle fatigue regime; this is approximately 235°C for 51200 steel and less than 80°C in 4240 steel;

two martensitic steels with and without retained austenite (Figures 3.2 and 3.5). Why is the difference between the thermal behavior of these two steels quite so important?

To answer this question, it is useful to confirm that the size, the shape and the diameter (3 mm) of the specimen are the same. The fatigue life is in the range of 10^8 cycles and the frequency of the test is 20 kHz for both cases. The piezoelectric machine and the CEDIP thermal camera are the same. In these conditions, it is assumed that the difference in thermal behavior is due to a microstructural effect. In both steels, the microstructure is martensite, but in the 51200 steel, depending on the thermal treatment, the level of residual austenite is high. This phase is not stable in cyclic loading. It is assumed that the transformation of austenite to martensite related to the local plastic deformation at the beginning of the test should explain the increase in temperature.

To sum up the observations concerning ultrasonic fatigue, several aspects of the microstructure stability are pointed out below:

– *Effect of austenite* in stainless steels: in austenic stainless steels, the austenite is not stable in the gigacycle fatigue regime even if the plastic deformation is theoretically very small. There is a large thermal dissipation. To keep the temperature low during testing, a strong cooling is necessary. However, the S-N curve is flat with a large scatter.

– *Effect of residual austenite* in martensite: when the rate of austenite approaches 10% in martensitic steels, a large thermal dissipation is observed at the beginning of the test followed by a high temperature when the fish eye propagates. The thermal dissipation associated with the microstructural instability is a limitation of ultrasonic fatigue. The phase transformation of martensite or bainite is not well understood in the face of the state of the art. More research is needed in this area. It would probably be a good way to understand why sometimes frequency has an effect on fatigue and sometimes it does not.

– *Effect of frequency* in ultrasonic fatigue: according to our results, the conclusion is that no significant effect is observed in gigacycle fatigue at high frequency if the temperature is kept close to room temperature, if no environment effect is operating and if the microstructure is stable. This means that it is of great significance to cool the specimen in ultrasonic fatigue in order to keep the temperature close to ambient.

– Relation between thermal dissipation and plastic deformation: a *strong coupling* is observed during the fish-eye growth. A thermal dissipation cannot be avoided during crack growth. This means that the significance of the piezoelectric fatigue systems must be limited, in particular, to the crack initiation in the gigacycle range and to the threshold crack regime up to the threshold corner of the Paris law.

3.6. Gigacycle fatigue of pure metals

It is clear that industrial alloys can fail in fatigue up to 10^{10} cycles, i.e. more than the fatigue life of many components. This means that during the service life of those components a failure can occur in the gigacycle regime. In this case, the concept of infinite life and fatigue limit is not efficient from a technological point of view. Such behavior is explained by the presence of defects, inclusions, pores or large grains in industrial alloys. However, gigacyclique fatigue also exists in pure metals such as copper or iron where the defects are very small. Some recent research on copper and iron have shown that failure can occur up to 10^9 cycles [WAN 13]. The phenomenon is same in both face-centered and body-centered cubic single-phase metals displaying planar or wavy slip. In high-cycle fatigue and very high cycle fatigue, the mechanisms are basically about the same. The irreversible microstructural changes are mainly related to dislocation processes such as persistent slip band (PSB) formation but in gigacyclique fatigue, the density of PSB is lower. This gigacyclique mechanism emphasizes the role of the surface in contrast to the industrial alloys. In copper, it has been found by Chrysochoos *et al.* [CHR 12] that the intrinsic thermal dissipation during initiation is not dependent on the frequency of the test. This tends to demonstrate that the key parameter in ultrasonic fatigue is not the frequency but the temperature. Confronting the technological significance of steel, the focus is on Armco iron in gigacyclique fatigue.

The very high cycle fatigue investigation was performed on a basic Armco iron. The material machined is in rolling condition with an original thickness of 1 mm. The ferrite single-phase microstructure has a grain size between 10 and 40 µm. The chemical composition is presented in Table 3.2.

C	P	Si	Mn	S	Cr	Ni	Mo	Cu	Sn	Fe
0.008	0.007	0.005	0.048	0.003	0.015	0.014	0.009	0.001	0.002	Balance

Table 3.2. *Chemical compositions (wt%)*

The monotonic tensile curve is shown in Figure 3.12. The upper yield point is 310 MPa and lower yield point is 250 MPa. The corrected section area before necking gives an approximate fracture stress at 400 MPa.

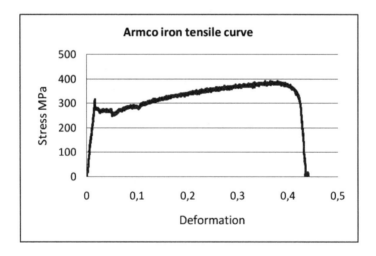

Figure 3.12. *Tensile curve for Armco iron*

The ultrasonic fatigue tests were carried out using a flat specimen, especially designed for microscopic observation and temperature measurement. All the tests were performed at 20 kHz. The first step was to study the effect of temperature on fatigue mechanisms and fatigue initiation in the gigacycle regime.

Thus, flat specimens are tested in different cooling conditions. Figure 3.13 shows the S-N curve for specimens tested without compressed cooling air in ambient air conditions. As the stress varies from 124 to 94 MPa, the fatigue life is in the range 3×10^6 to 4×10^9 cycles. A slope may be drawn as shown in the figure. The lifetime is increased by lowering the fatigue stress amplitude. Pronounced irreversible slip bands appear on the specimen surface as discussed in Chapter 2. It is of significance to note that Armco iron failed in fatigue beyond 10^9 cycles. Indeed, there is no fatigue limit in iron at 10^6 cycles as predicted by the conventional standard.

At a test frequency of 20 kHz, the experiment shows that the temperature of the plateau T_0 is between 65°C and 85°C in the gigacyclique regime.

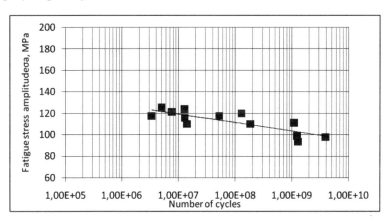

Figure 3.13. *S-N curve of flat specimen without cooling [WAN 13]*

The S-N curve of specimens tested in 5 bar cooling air is shown in Figure 3.14. Contrary to ambient conditions shown above, a 5 bar air cooling surface decreases the irreversible slip band (but existing) formation. It is suggested that appearance of irreversible slip bands is strongly dependent on cooling conditions during ultrasonic fatigue testing. Again, the Armco iron failed beyond 10^9 cycles.

At a test frequency of 20 kHz, with a 5 bar air cooling, the temperature T_0 is between 32°C and 38°C.

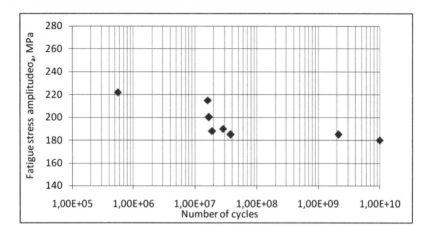

Figure 3.14. *S-N curve of flat specimen cooling in 5 bar*

The last series of flat specimens was also tested in cooling air but with lower pressure (2 bar). The S-N curve in Figure 3.15 shows the results. Fatigue life changes from 10^5 to 10^{10} in a narrow variation of stress at approximately 12 MPa.

With an air cooling pressure of 2 bar, the temperature of the plateau T_0 is between 50°C and 60°C.

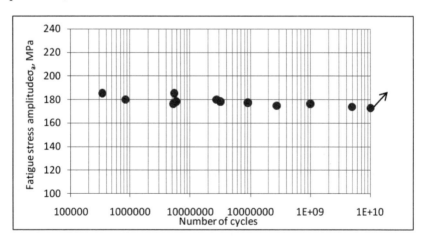

Figure 3.15. *S-N curve of flat specimen cooling in 2 bar*

All results of flat Armco iron specimens are shown in Figure 3.16. It is found that the gigacycle fatigue property is significantly influenced by the cooling condition (ΔT in the range 12–65°C). When fatigue tests are carried out with cooling air, the fatigue stress amplitude is approximately 60 MPa more than the tests carried out without cooling air.

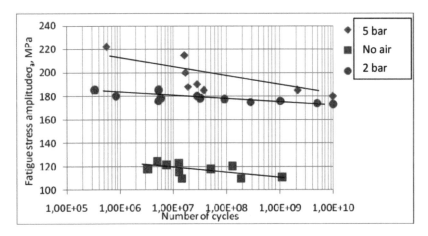

Figure 3.16. *S-N curve of flat Armco iron specimen in different cooling conditions [WAN 13]*

From those results, it can be concluded here that in very high cycle fatigue tests, the fatigue life is significantly affected by the testing temperature, more than by the high frequency.

3.6.1. *Microplasticity in the ferrite*

At the microscale level, Mughrabi [MUG 06] suggests that the initiation of fatigue crack in the very high cycle fatigue (VHCF) regime can be explained by the PSB formation due to the irreversible portion of the cumulative cycle strain. It has been concluded that intrusion and extrusion lead to the crack initiation in very high cycle fatigue. However, there is still a sufficient lack of literature which studies the irreversible deformation on the surface for the base centered cubic (BCC) crystalline in the very high cycle regime.

The work of Hempel [HEM 59] also indicates the results ($\Delta\sigma$ against cyclic number N) when the first PSB occurs during a fatigue test, as shown in Figure 3.17. It is found that in a ferrite specimen, at a stress range 162–245 MPa, first PSB occurs on the surface after 2,400–2.5 × 10^4 cycles. The difference when the first PSB takes place is approximately 10 times between 162 and 245 MPa. This is regular scatter. It means that the appearance of the first PSB is independent of N_f. However, the difference of occurrence of fatigue failure is in the order of 1,000 times. In other words, at a low stress level, the first PSB occurs at the beginning of fatigue test, but needs much more repeated loading to reach the failure of the specimen.

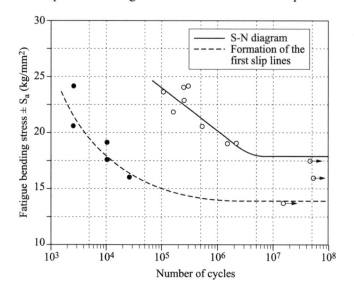

Figure 3.17. *Occurrence of the first slip line on unnotched flat specimens of 0.09% C steel [HEM 59]*

Through the advantages of displacement nodes and stress distribution at the center of ultrasonic fatigue, the specimen is under online optical microscope observation. It is found that irreversible slip bands are formed at the beginning of fatigue loading. These slip bands are multiplied or grown in the same grain due to the following of loading. It can result in new slip bands. At the slip bands' site, where optical observation was carried out, most of the irreversible slip bands formed before 3×10^8 cycles. Only very slight evolution occurs by fatigue loading during cycles between 3×10^8 and 10^{10} cycles that are newly irreversible. In other words, crack initiation due to cumulating further new irreversible deformation after 3×10^8 cycles is scattered. This may be the reason why the S-N curve is significantly dispersed when failure takes place in VHCF regime.

3.6.2. *Effect of gigacycle fatigue loading on the yield stress in Armco iron*

Tensile tests were carried out on fatigue specimens before and after fatigue loading (2 bar cooling air). Deviation of mechanical properties occurs between results with and without gigacycle fatigue loading, as shown in Figure 3.18. Yield stress of the specimen is decreased after ultrasonic fatigue loading.

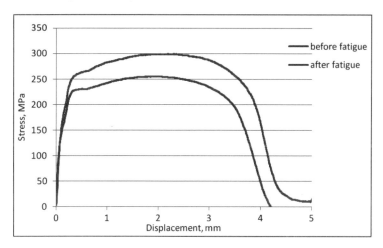

Figure 3.18. *Tensile curve for a flat specimen before and after ultrasonic fatigue (σ_a: 173 MPa, N: 5×10^8 cycles). For a color version of the figure, see www.iste.co.uk/bathias/Fatigue.zip*

The decrease in yield stress after fatigue loading shows that even in the VHCF domain, a modification of the dislocation pattern occurred, which shows that even at a fatigue stress amplitude below the yield point, a glide of the dislocation is possible leading to a microplasticity. At the same time, a reorganization of the solute atoms (C, N) probably occurs. The anchorage of dislocations is not so strong and the yield stress is decreased. This phenomenon plays a role in the temperature increase T_0 during gigacycle testing.

3.6.3. *Temperature measurement on Armco iron*

Through the advantage of online microscope observation, the surface microplasticity evolution has been investigated during uninterrupted gigacycle fatigue testing and compared with the temperature measurement. The occurrence and evolution of microplasticity on the surface of a ferrite specimen is important in order to analyze the coupling of plasticity and thermodynamics. Fractographic analysis on the fatigue crack surface shows that microplasticity at the surface and subsurface of a ferrite specimen leads to the fatigue initiation. The threshold of fatigue crack propagation in a flat ferrite specimen is determined according to the Herzberg model [HER 98]. A combined model of fatigue crack initiation and propagation is discussed according to thermography, fracture analysis and microscopic observation.

An FLIR IR camera was used to measure the thermal field during gigacycle fatigue. For example, the thermal recordings obtained on some specimens are reported below.

Figure 3.19 shows the temperature evolution against lifetime for a flat specimen. This curve is same for flat and round specimens, during fatigue testing from the beginning until failure, characterized in three major stages. After the first increase in the temperature and a long plateau, the slope of the curve increases drastically from 35°C up to 250°C and even higher at the end of this test. The test detail is as follows:

Parameter of IR camera	Specimen: FF325
Resolution of detector: $130 \pm 10 \ \mu m^2$/Pixel	$R = -1$
Acquisition frequency: 100 Hz	$\sigma a = 175$ MPa
Spectral domain: 2.0–4.8 μm	$Nt = 3.255 \times 10^7$ cycles
Integration time: 100 μs	Cooling air: 5°C, 2 bar

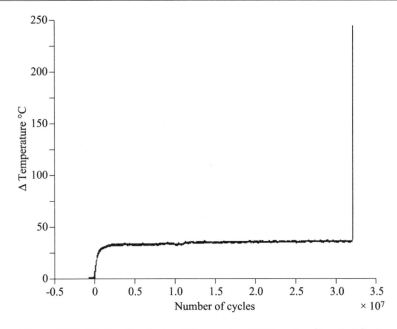

Figure 3.19. *ΔT elevation during fatigue test until failure (cooling air 2 bar)*

Typical ΔT profiles along the length of the flat specimen for the different stress amplitudes are shown in Figure 3.20. The maximum temperature occurs at the specimen's center; the maximum stress is induced by the steady wave. However, the profile is not perfectly symmetrical due to the difference in thermal conductivity at the boundary condition (one side is connected to the acoustic horn and the other side is free). As a result, the fatigue crack section is slightly shifted from the center (maximum stress) to the side where the temperature is higher and the PSB density is greater.

We find that increasing the stress amplitude of the cyclic loading increases the temperature of the plateau expressed by ΔT or T_0. Figures 3.21 and 3.22 give ΔT evolution as a function of the number of cycles at different stress amplitudes when the specimen is cooled by fresh air and when the specimen is not cooled. Both figures indicate that the plateau of temperature occurs (variation < 5%) after 3×10^6 cycles from the beginning of the test.

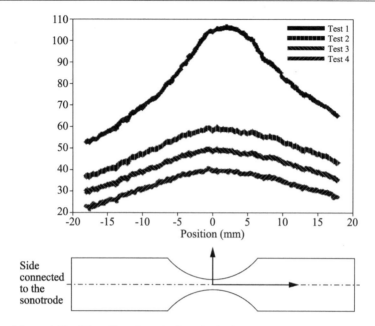

Figure 3.20. *ΔT profiles along the length of a flat specimen for several loadings. For a color version of the figure, see www.iste.co.uk/bathias/Fatigue.zip*

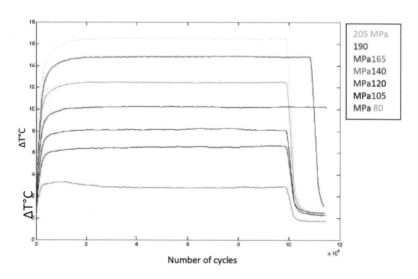

Figure 3.21. *ΔT elevation curve (cooling in 5 bar), R = −1. For a color version of the figure, see www.iste.co.uk/bathias/Fatigue.zip*

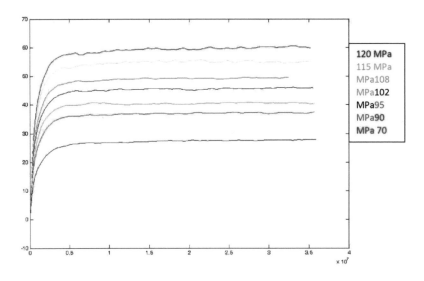

Figure 3.22. *ΔT increase curve (no cooling). For a color version of the figure, see www.iste.co.uk/bathias/Fatigue.zip*

Comparing stable ΔT at different stresses and different cooling conditions, we find that at each cooling condition, stable ΔT is in linear relation to the stress amplitude in the VHCF domain. However, cooling condition affects the slope of the linear relationship between stable ΔT and stress as shown in Figure 3.22. It is found that when this linear relationship to the ΔT = 0 is extended, the fatigue stress amplitude does not go down to 0 but intersect at approximately σ_a=50 MPa (Figure 3.23). Considering the stress concentration at grain boundary for polycrystalline material and experimental yield stress for single crystals iron $\sigma_{exp} \approx 2 \cdot \tau_{exp} = 2 \times 27.5 = 55$ MPa, it is suggested that at ΔT = 0 for different cooling conditions, approximately $\sigma_a \approx \sigma_{exp}$.

From this observation, the extrapolation of the relation between temperature increase and cyclic stress show that below 50 MPa, the temperature increase is close to zero. It means that a limit for the PSB formation should exist in Armco iron. Therefore, from a fundamental point of view, a fatigue limit should exist. However, the threshold of 50 MPa is so small that its technological meaning is questionable.

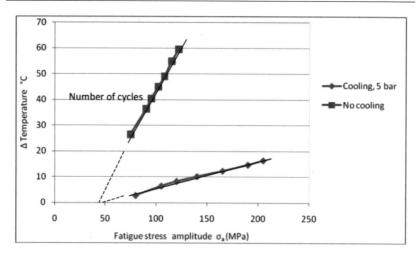

Figure 3.23. *Estimation of the amplitude of the stress
for no heating dissipation*

3.6.4. *Intrinsic thermal dissipation in Armco iron*

The average intrinsic dissipation along the width of the specimen was investigated for different locations on the length of a flat Armco specimen the failed in ultrasonic fatigue [BLA 12]. Profiles of this average intrinsic dissipation are shown in Figure 3.24 at different periods of fatigue testing. At the beginning of the test, the average section intrinsic dissipation is very low (<6°C/s). In this test, since total fatigue life is less than 10^7, there is no stable plateau as shown in Figure 3.19. The average section intrinsic dissipation was raised continually and slightly by following cycles. Until a few seconds before the fatigue failure, the average section intrinsic dissipation was raised rapidly in a short time. For the reason of localized stress concentration, which may be caused by grain disorientation and axial loading [MCD 10], the gigacycle fatigue failure frequently occurred asymmetrically at the center section of the specimen. Maximum intrinsic dissipation along the specimen length at the final stage of fatigue indicates well the section where the crack takes place, as shown in Figure 3.24.

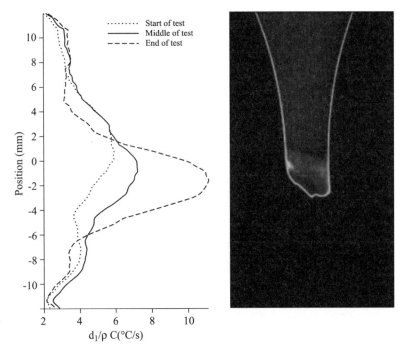

Figure 3.24. *Fer Armco 124 MPa, $N_t = 5 \times 10^6$ cycles. Intrinsic dissipation from the beginning to the failure [BLA 13]. For a color version of the figure, see www.iste.co.uk/bathias/Fatigue.zip*

Therefore, the position of maximum intrinsic dissipation is reliable and varied. Because of the maximum dislocation slips, density is in the center of specimen at the beginning of the test. However, it may move slightly due to the shear band orientation in the plane stress condition and with the asymmetry of the temperature in the flat specimen.

Intrinsic dissipation in a two-dimensional (2D) map gives more information to relate the energy dissipation to the fatigue failure (Figure 3.25). The top of Figure 3.24 shows the temperature recording on the specimen surface against the number of cycles. The maps for different numbers of cycles are related to the temperature recording. The map's abscissa corresponds to the specimen width and the ordinate corresponds to the specimen length at the bottom of Figure 3.24.

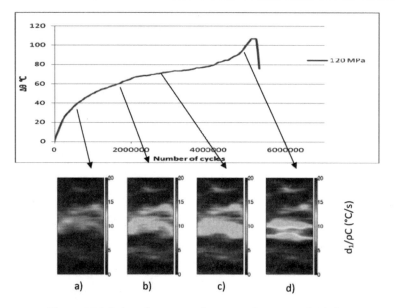

Figure 3.25. *Infrared images and intrinsic dissipation fields for different numbers of loading cycles [WAN 13]. For a color version of the figure, see www.iste.co.uk/bathias/Fatigue.zip*

It is found that, as discussed above, the intrinsic dissipation increases with increasing fatigue cycles. The maximum dissipation appeared on two sides in the middle of the specimen. It is in quite good agreement with the stress concentration along the hyperbolic specimen. At the beginning and in the middle period of the fatigue life, intrinsic dissipation is small on the entire surface. Finally, the dissipation reaches 20°C/s.

At the cross-section in the middle of the specimen, the intrinsic dissipation is the highest on the right side, which corresponds to the crack initiation site in Figure 3.25 (upper part of failed specimen). Around the initiation area on the specimen surface, irreversible slip bands were found that may correspond to the lower level intrinsic dissipation before the fatigue crack takes place. Note that the intrinsic dissipation is the trace of the specific heat sources (or, in other words, plastic deformation) in the material that is subjected to fatigue loading. It shows the drastic effect of temperature on initiation and PSB formation.

Furthermore, SEM observations have been performed on the polished surface where the temperature was recorded. The initiation stage (stage I) corresponds to the highest dissipation d_I on the right side of the last dissipation image as shown in Figure 3.5. The total length of both sides of fracture surface where PSBs occurred is approximately 6.3 mm, in other words, it is in very good agreement with the length where d_I is highest.

3.6.5. *Analysis of surface fatigue crack on iron*

Intrinsic energy dissipation is an excellent parameter for the calculation of the energy balance between energy stored in metal by plastic deformation and energy for nucleation and propagation of a small crack. However, the maximum dissipation is always localized at the location where the maximum temperature occurred. Therefore, in the first approximation, there is a good correlation between the tip of the crack during propagation and the location of the maximum temperature at the surface of a flat specimen. The specimen in Figure 3.26 gives an example where the maximum temperature occurs at the tip of the crack. There is no basic difference between a flat specimen and a round specimen, but it is much easier to measure the temperature on a flat surface and at the tip of a part through a crack. Indeed, it is more difficult when the crack is growing beneath the surface or in the interior.

During a test carried out under the 2 bar cooling air (Figure 3.27), the temperature recording on the specimen surface versus the number of cycles clearly shows the difference of heating dissipation before and after the beginning of the crack propagation. Once again, more information is obtained from a flat specimen than from a round specimen. Figure 3.27 shows the temperature increase in the width of a flat specimen from the end of the initiation stage to the crack instability. The temperature profile along the specimen width was extracted for the latest 168 pictures captured by the camera, i.e. means from 3.2086×10^7 cycles to 3.2153×10^7 cycles (number of the cycles at the failure). For more than 3×10^7 cycles, the temperature across the specimen is almost constant (35–50°C), focused on points 1 and 2. From point 2 to point 6, the temperature increases from 50°C to 500°C very quickly, in 8.7×10^4 cycles.

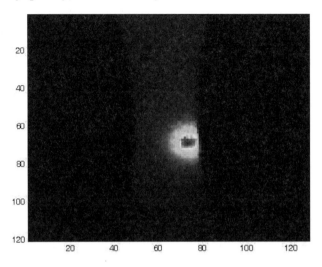

Figure 3.26. *Thermograph at a crack tip in a flat specimen. For a color version of the figure, see www.iste.co.uk/bathias/Fatigue.zip*

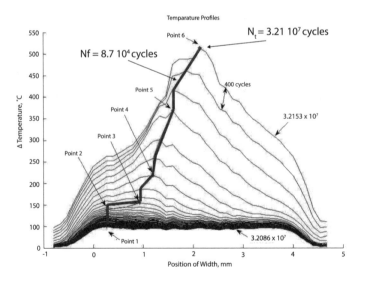

Figure 3.27. *Temperature increase along the width of the specimen from initiation to failure. For a color version of the figure, see www.iste.co.uk/bathias/Fatigue.zip*

Between each step of the temperature profile in Figure 3.27, there are 400 cycles. The bold line (red in the color version) shows the location where the temperature is maximum along the profile. It must be pointed out that the temperature is almost constant through the width of the flat specimen before 3.2086×10^7 cycles, but after initiation, the maximum of the recorded temperature is always on the same side of the specimen from which the crack is growing. At the end of the test, the location of maximum temperature moves through the specimen width according to the fatigue crack length and the increase in the plastic zone ahead of the fatigue crack tip.

The distance between the points 1/2, 3, 4, 5 and 6 is determined from the specimen corner where the initiation of the crack occurs: points 1 and 2 is 267 μm, point 3 is 933 μm, point 4 is 1,200 μm, point 5 is 1,600 μm and point 6 is 2,000 μm. The corresponding numbers of cycles are: N_1 (point 1): 3.2086×10^7 cycles, N_2 (point 2): 3.2148×10^7 cycles, N_3 (point 3): 3.2149×10^7 cycles, N_4 (point 4): 3.2151×10^7 cycles, N_5 (point 5): 3.2152×10^7 cycles, N_2 (point 2): 3.2153×10^7 cycles and the number of cycles at fracture N_t is 3.2155×10^7 cycles. Roughly speaking, it is concluded that the fatigue life for initiation is of the order of 3.2×10^7 cycles since the fatigue life for propagation is approximately 8.7×10^4 cycles. These results confirm that the initiation of the crack is the key problem in a flat specimen that failed in gigacycle fatigue, whatever the location of the initiation, surface or interior. The conclusion is the same as that for a round specimen.

From the macroscopic observation of the crack at the surface of a flat specimen, it is possible to determine its shape and size and calculate the stress intensity factor. According to the Paris–Hertzberg model, it is assumed that the transition between initiation and propagation is given by the threshold corner. Thus, it is of significance to calculate the stress intensity factor at the point 1 or the point 2 as shown in Figure 3.28, typical of gigacycle fatigue in an Armco iron flat specimen.

The ΔK_{eff} was calculated using the relation:

$$\Delta K_{\mathrm{eff}}|^{\mathrm{ai}} = Y \cdot \Delta \sigma \cdot a^{1/2}$$

where a is the crack length and Y is a parameter corresponding to the crack geometry. For this specimen, the fatigue stress intensity factor threshold is calculated using the effective fatigue stress intensity factor (between initiation and propagation). It is found to be $3 \text{ MPa·m}^{1/2}$.

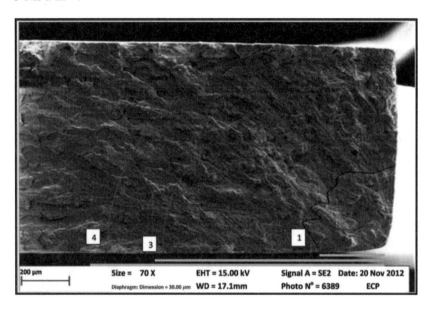

Figure 3.28. *Fracture surface showing three stages at points 1 (2), 3 and 4*

According the Paris-Hertzberg model, the theoretical threshold for iron is given by:

$$\Delta K_{th} \approx \Delta K_{eff}|^{ao} = 3.98 \text{ MPa·m}^{1/2}, \text{ where } a_o \text{ is crack length (point 1/2)}.$$

Therefore, the stress intensity factor becomes greater than the threshold beyond point 1. It is a confirmation of the meaning of point 1: transition between initiation and propagation. It is also the confirmation of the large number of cycles required to initiate the crack, whatever the mechanism, fish-eye or PSB formation, at the surface in gigacycle fatigue.

3.7. Conclusion

According to recent experimental results obtained during the 2000s, it is obvious that gigacycle fatigue exists and the concept of infinite fatigue life is not correct, at least for practical applications. Several conclusions can be pointed out:

– It has been found that a coupling exists between thermal dissipation, plasticity and damage in gigacycle fatigue regime. Thermal dissipation helps to characterize propagation, initiation and PSB formation.

– From a fundamental point of view, it is difficult to conclude because it seems that a threshold for the PSB formation would exist. The accuracy of the microscopic observation and of the thermal measurement does not allow us to reach a conclusion.

– Assuming that the alloys can fail beyond 10^{10} cycles and that the most sophisticated mechanical components are designed for a fatigue life more or less in the order of the gigacycle regime, the standard megacycle fatigue limit is not conservative. Thus, the significance of a gigacycle curve is recommended. A new standard for the fatigue strength is mandatory.

– The thermal dissipation and fracture mechanics approach demonstrate that the initiation of a crack is the crucial mechanism in gigacycle fatigue.

– Thermal dissipation is a good non-destructive technique to detect damage in gigacycle fatigue even if initiation is beneath the surface. However, the damage tolerance concept is difficult to apply.

– The temperature increase during an ultrasonic test depends on the microstructure of metal, the phase transformation, if any, the anelasticity, the microplasticity and the plastic zone at the crack tip.

– To get a good agreement between a conventional fatigue test and an ultrasonic test, it is recommended to cool the piezoelectric machine with compressed fresh air, keeping the specimen at room temperature.

3.8. Bibliography

[BAT 10] BATHIAS C., *Fatigue of Materials and Structures*, pp. 311–373, John Wiley and Sons, New York, 2010.

[BAT 13] BATHIAS C., "Coupling effect of plasticity, thermal dissipation an metallurgical stability in ultrasonic fatigue", *International Journal of Fatigue*, 2013.

[BLA 12] BLANCHE A., Effets dissipatifs en fatigue à très grand nombre de cycles, PhD Thesis, University of Montpellier, 2012.

[CHR 00] CHRYSOCHOOS A., LOUCHE H., "An infrared image processing to analyse to analyse the calorific effects accompanying strain localization", *International Journal of Engineering Science*, vol. 38, no. 16, pp. 1759–1788, 2000.

[CHR 12] CHRYSOCHOOS, Private communication, DISFAT Report 2012 ANR. (French Agency for Research)

[HEM 59] HEMPEL M.R., "Slip bands, twins, and precipitation processes in fatigue stressing", *Fracture*, vol. 19, pp. 376–411, 1959.

[HER 13] HERVE P., Physique fondamentale, Thermique, Handbook, Les Techniques de l'Ingénieur, 2013.

[LUO 98] LUONG M.P., "Fatigue limit evaluation of metals using an infrared thermographic technique", *Mechanics of Materials*, vol. 28, no. 1, pp.155–163, 1998.

[MOO 21] MOORE H.F., KOMMERS J.B., "Fatigue of metals under repeated stress", *Chemical and Metallurgical Emgineering*, vol. 25, pp. 1141–1144, December 1921.

[RAN 10] RANC N., WAGNER D., PARIS P.C., "Study of thermal effects associated with crack propagation during very high cycle fatigue", *Acta Materiala*, vol. 56, no.15, pp. 4012–4021, 2010.

[WAG 01] WAGNER D., RANC N., BATHIAS C., *et al.*, "Fatigue crack initiation detection by an infrared thermography method", *Fatigue & Fracture of Engineering Materials & Structures*, vol. 33, no. 1, pp. 12–21, 2001.

[WAN 13] WANG CHONG, Microplasticity and dissipation in VHCF of iron and steel, PhD Thesis, University of Paris, 2013.

Index